S0-ATF-885

Fear of Math

Fear of Math

HOW TO GET OVER IT AND GET ON WITH YOUR LIFE

Claudia Zaslavsky

Rutgers University Press
New Brunswick, New Jersey

For permission to publish their cartoons, the author thanks the following:
Page 83: North Dakota Department of Public Instruction; ND State Board for Vocational Education; ND Governor's Council on Human Resources.
Page 98: David Shenton, artist; it first appeared in *The Guardian* (U.K.) (15 March 1989).
Page 108: by Wasserman. Copyright 1981, *Boston Globe*. Distributed by Los Angeles Times Syndicate. Reprinted by permission.
Page 130: William T. Coulter, artist.
Page 189: Copyright 1991 EQUALS, Lawrence Hall of Science, University of California, Berkeley.
Page 191: from *Mathematics: The Invisible Filter*, by the Mathematics Department, Board of Education for City of Toronto.
Page 211: Copyright 1988 by the Regents of the University of California; FAMILY MATH, Lawrence Hall of Science.
Pages 62, 139, 148, 180, 203, 205: Karen Reeds for her cartoons.
For permission to use their photographs, the author thanks the following:
Page 119: The *New York Teacher* and Mina Choi.
Page 134: Donald W. Crowe.
All other photographs and illustrations are by Sam Zaslavsky.
For permission to reprint the following selections, the author thanks:
Page 50: Phyllis R. Steinmann for "Another Longitudinal Study."
Page 144: Gary Hendren and the MCTM *Bulletin* for the word problem.
Page 166: Beverly Slapin and New Society Publishers for "Two Plus Two or Why Indians Flunk."

Library of Congress Cataloging-in-Publication Data

Zaslavsky, Claudia.
Fear of math : how to get over it and get on with your life / Claudia Zaslavsky.
p. cm.
Includes bibliographical references and index.
ISBN 0-8135-2090-8 (cloth)—0-8135-2099-1 (pbk.)
1. Math anxiety. 2. Mathematics—Study and teaching.
3. Self-help techniques. I. Title.
QA11.Z37 1994
370.15'651—dc20 93-43904
CIP

British Cataloging-in-Publication information available

Manufactured in the United States of America

To Sam, for unfailing help with this book,
and in many other ways

Contents

Acknowledgments

This book is truly the outcome of a joint effort. The initial impetus came from my friend Rose Wyler. She had the foresight to predict that such a book would be both useful and necessary. I hope she is right.

I want to acknowledge the many women and men who gave of their time, effort, and expertise to make this book possible. They responded to my inquiries by phone, mail, and personal conversation. They sent articles, books, and unpublished papers. They furnished valuable information about resources, reports, and meetings, and made further contacts possible. Many of these people are cited in the text and the notes. To all, I extend my appreciation for their valuable assistance.

I am grateful to the hundreds of people who took the time and trouble to write their math autobiographies at my request. Their experiences with mathematics were often inspiring, but more frequently devastating. I would have liked to include all of them. To those contributors named in the text, and to the many more identified only by assumed first names, I say a heartfelt "Thank you." I owe a special expression of gratitude to Mary Jo Cittadino, whose inspiring words I have used throughout the book.

Perhaps as difficult as writing a math autobiography was the task of collecting such information from others—relatives, friends, associates, and students. For their tremendous assistance I want to thank Barbara Barone, Joanne Rossi Becker, Sylvia Bozeman, Manon Charboneau, Gil Cuevas, Linda Falstein, Len Feldman, Maryam Hastings, Don Hill, Lotus Jones, Pat Kenschaft, Vera Preston, Gayle L. Smith, Phyllis Steinmann, Lyn Taylor, Marion Walter, Betsey Whitman, Pat Woodruff, and Suzette Wright.

Several experts read sections of the manuscript in its various drafts and offered valuable critiques, advice, and materials. For this time-consuming contribution I want to thank Olivia

Abelson, Jon Beckwith, Lotus Jones, Frances McGhee, John McGhee, Lucy Sells, Ethel Tobach, Joe Washington, Suzette Wright, Alan Zaslavsky, Sam Zaslavsky, and especially Pat Kenschaft.

I am indebted to my editors: to Pat Woodruff for getting me started, to Miriam Schocken and Susan Rabiner, and finally, to Karen Reeds of Rutgers University Press for her faith in the project, for her many constructive criticisms, for her cartoon art, and for seeing the book through to the end.

Very special thanks are due my family—my husband, Sam, who did the line drawings; my sons Tom and Alan, and their wives, Zabeda and Noel—for their never-failing support and their good suggestions, based on their own relevant experiences.

Fear of Math

Introduction

This book has a vast potential audience. It is addressed to all of you who have negative feelings about mathematics, whatever the reasons. My main motive for writing this book is to show you that you are not to blame for those negative feelings. Most likely you are a victim of the many conditions in our society that bring about fear and avoidance of math. Do not despair! As many people have already found out, it is possible to change feelings about mathematics from negative to positive.

Who are all you math-fearing and math-avoiding people?

- You may feel inadequate because math never did make sense to you. It is possible to overcome those feelings of inadequacy.
- You may be a college student, whether a recent high school graduate or an older adult going back to school after many years. You can learn to cope with the required math courses.
- You may be a worker enrolled in a training program to upgrade your skills for the new technologies. You can learn to handle those unfamiliar math problems.
- You may be a parent, teacher, or caregiver of young children. You don't want to pass along your negative attitude to the next generation.

This book offers help to those of you who need it right now. It includes excerpts from dozens of "math autobiographies," some written by math fearers, others by those who have overcome their fear and avoidance. They are people with whom you can identify and say, "That's how I feel" or "That's how my problem started." Pinpointing the problem is a first step in overcoming negative feelings about math.

I will discuss the many factors in our society that may have led to your difficulties with mathematics. Among them are inadequate schools, poor teaching, inappropriate mathematics programs, and stereotypes about who can and who should do math.

INTRODUCTION

In the following chapters I offer concrete suggestions to help you attain mathematical competence, based on the experiences of successful teachers and learners. I recommend enjoyable activities you can use to give young children an early start in math, regardless of your own mathematical inadequacies.

The book includes real-life biographies of people who overcame great odds to achieve success in math-related fields. It gives examples of the mathematical practices of many cultures. It includes real-life problems to help you as a concerned citizen to make informed decisions about important issues in our society.

This is not the first book to deal with the subject of "fear of math" or "math anxiety." But I believe that this is the first book for the general public to focus on the many factors in our society that bring about fear and avoidance of mathematics, and the influence of these factors on various groups in the population. Previous treatments of the subject have centered on the psychological effects of math phobia, with the goal of encouraging "math-anxious" individuals to overcome their fears and inhibitions by self-help methods.

The problem of "math anxiety" came to the fore in the mid-1970s as a feminist issue. Young women were not taking the high school math courses they needed for many college majors, and, as a consequence, they were excluding themselves from promising and well-paying careers. Lucy Sells, a California sociologist, called mathematics the "critical filter." Some colleges responded by setting up clinics to serve their math-avoiding female students. Most of the books, articles, and research studies dealing with fear and avoidance of mathematics have addressed the issue of gender and mathematics. Incidentally, some women in the field have objected to the use of the term "anxiety" as implying a form of psychological illness, rather than a condition that can be overcome.

A far greater problem of math avoidance is the exclusion of many working-class people and people of color from the growing number of careers requiring some mathematical background, or, at the very least, an entry test on topics in mathematics. Moreover, most colleges now have mathematics requirements for all students. There is a growing awareness of the need for studies exploring the interconnections of race and ethnicity with economic and social status as they affect mathematics achievement and restrict access to college education and careers.

INTRODUCTION

In this book I have drawn upon my many years of experience in teaching mathematics at different levels—to teachers, aspiring teachers, secondary and elementary school students. I was fortunate in having taught secondary school mathematics in a New York State district that mandated integration by busing in 1951, three years before the famous Supreme Court decision on school integration. All our elementary school classes were balanced for ethnicity, gender, and socioeconomic level. Although young women were well represented in advanced high school mathematics courses, few African Americans continued to that level. When we wrote our own courses, more friendly and relevant than the standard curriculum, many more students, both black and white, studied mathematics all through high school.

As you read this book you will hear the voices of people from different ethnic/racial backgrounds, ranging in age from fourteen to sixty-eight. In response to my request I received over five hundred math autobiographies—from friends and acquaintances, from their relatives and friends, and from instructors of a diversity of students. It was not easy to select representative samples to include in this book. More women than men were willing to express their feelings and doubts. Although I have changed the names when requested, I am sure that the writers will recognize themselves. These anonymous voices tell their experiences with learning and doing math. I have also cited research, wherever available, to back up their stories. One unpleasant incident in third grade really can rob a young woman of confidence in her ability to take a college course in statistics. Sensational, and unfounded, newspaper reports about boys, but not girls, having a "gene for math" really do influence parents and teachers. Placement in the "slow track" in second grade really does mean that African-American children or non-English-speaking children may not be eligible to study algebra and other college entrance math courses in high school.

The question of assigning labels to ethnic and cultural groups is a difficult one to resolve. A name that is acceptable to some people may be anathema to others. In the media and in many government reports, groups that are not "white" are frequently lumped together as "nonwhite" or "minority." "Nonwhite" implies that "white" is the standard, like defining women as "non-men." The word "minority" may carry the implication of a lower status than the "white" majority. Furthermore, "minority" lumps together many diverse groups as though they are all the same.

There are problems, too, with more specific names. There is disagreement about the use of black, Black, or African American, about Latino/Latina rather than Hispanic. Each of these groupings covers people of varied backgrounds. African Americans include those born in the United States, in Caribbean countries, and elsewhere, of different economic levels, and from urban, suburban, and rural backgrounds. Hispanics (people with Spanish-language background) may be mainland or island Puerto Ricans, of Mexican origin, from South or Central America or the Caribbean, and from diverse social, racial, and economic backgrounds. A similar analysis applies to Asian Americans. The term "Native American" is preferred by some indigenous people, while others would rather be considered American Indian, but all agree that they prefer to be identified by their ethnic group.

How to resolve this problem? At times I have used the terms that appear in the media and in many research reports: minority, black, white, Hispanic. At other times I have used different designations. To distinguish between the dominant "white" majority and a group that includes African Americans, Asian Americans, Latinos/Latinas, and American Indians, I have also used the term "people of color."

In the final analysis, I hope that you will look beyond the labels and accept this book for its discussion of urgent societal problems that affect so many of us.

CHAPTER ONE

Who's Afraid of Math?

> Almost immediately following graduation [from high school], I found not understanding mathematics, not being able to *do* math, a problem. Four years of no math or science contributed to a gradual decline in my confidence—not only about math but about intelligence. It didn't happen all at once, but it happened inexorably.
>
> As a result, I avoided any confrontation or questioning in normal situations of making any kind of purchase. It was too scary. I also didn't even always know to question, or how to question. I needed to go away by myself and figure it out with paper and pencil to determine if a mistake had been made.
>
> —*Mary Jo Cittadino, Mathematics Educator, EQUALS, University of California at Berkeley*

Many people think of mathematics as one of the most logical, most impersonal branches of knowledge, yet it inspires more emotion than any other school subject. Whenever I mention to any group of people that I am writing a book to help those who fear or avoid math, immediately someone will say: "I never could do math!" or volunteer an experience that led to fear or avoidance. It happens with even the most casual acquaintances—the lawyer sitting next to me at a concert, the young woman on the ski lift. In the United States most people would be ashamed to admit that they never could learn to read, yet it is perfectly respectable to confess that one can't do math.

Mathematics has such a bad reputation in this country that it can be used to *induce* emotional stress. Physicians measuring blood flow under various stress conditions give their patients "a barrage of mental arithmetic problems" as a surefire method of inducing stress.[1]

Many people recall math as punishment. The humorist Russell Baker remembered how his seventh-grade class was to have the privilege of hearing a broadcast of a live concert. Any student who preferred not to listen was excused to go to another classroom. No fan of "longhair music," Baker gladly left with his pals. "We rebels had contemplated a pleasant hour pelting each other with erasers and upsetting ink jars. Imagine then our despair to see stern Miss Ward, the teacher who struck fear in the hearts of boys, stride into the room, command absolute silence and hand out thick wads of arithmetic problems to kill the hour."[2]

Math anxiety can afflict the most competent and intelligent people. I always enjoy listening to Steve Post's "Morning Music" program on New York's public radio station. One morning he read excerpts from a newspaper article about an experiment in which a small group of exceptionally nervous high school students had been given a drug to induce relaxation (not recommended for general use!) before retaking the Scholastic Aptitude Test (SAT). Although students who take the test over again without special preparation typically increase their math scores by an average of twenty points, these students improved by seventy points.[3]

This bright, perceptive, intelligent commentator went on to tell of his own experiences in the academic world. His school performance had been so poor that he had never been considered eligible to take the SATs. Math was his worst subject, and he recalled with dread the nightmare before a mathematics test: the test was written on his bed sheet, but because he was lying on it, he couldn't even read the questions, much less answer them. He claimed that the fear had stayed with him to that day.[4]

Negative feelings about math can take a variety of forms. Some people suffer from fear of math, while others have simply avoided the subject. Perhaps lack of confidence in their mathematical ability was at the root of their avoidance, or they were just not interested—or a dozen other reasons. There are the "weak-in-math" and the "rusty-in-math" individuals. Some fit into more than one category. Any one of these conditions may be involved in the syndrome known as "math anxiety" or "mathephobia," a state of mind that makes it difficult or even impossible for you to use the math skills that you already have. It may be accompanied by physical symptoms, such as headaches, nausea, heart palpitations, and dizziness. It surfaces in its most extreme form when you have to take a test. Although a certain amount of anxiety about a new or stressful situation is perfectly normal, and may even stimulate you to perform well, the term "math anxiety" is reserved for a condition that is traumatic and debilitating.

Who are these math fearers and avoiders? They are people of all ages, and from all walks of life. They are highly placed professionals and high school dropouts. They are at all levels of mathematical incompetence, from simple arithmetic to advanced college courses. They are white, black, Asian, Latino/Latina, American Indian, of all ethnic and racial backgrounds. And, contrary to the general belief, they are male as well as female.

Originally math anxiety was considered an affliction peculiar to women, but as Steve Post's experience illustrates, men are also subject to this syndrome. Here is an entry from the journal of an adult male college student:

> Growing up, I was raised by my grandparents who had definite roles (i.e., man breadwinner—woman housewife), so I was conditioned to a lot of sexist stereotypes. Math fits into that mold. I feel that my fears of math, and thinking that I could never do it, have always been a subtle reflection on my masculinity. Not that I ever doubted being a man, but maybe a little less of a man when I would stop reading an article or report when the author started throwing statistics or other numbers around.[5]

A teacher of adults in London had this to say about the male students in her statistics class: "I think the men were suffering more than the women, because at least the women were able to express their frustration and fears. Having done so, they then got on with the test and said: 'Oh, it wasn't so bad after all. I can do it.' " She cited the case of a man in the class who was unable to put his pen to the paper to carry out an assignment. "All I can see is that bastard, my father, standing over me," he muttered through clenched teeth.[6]

As I thought about this book, I started to gather research reports and news articles concerning math fear and avoidance. But more than reports and statistics, I wanted the living stories of those who were affected. I wanted you, the reader of this book, to know that you are not alone, that there are millions out there who also fear and shun mathematics. I started by speaking to friends and acquaintances. Then I asked several college math instructors in various parts of the country to have their students write about their experiences and feelings. One professor sent me over two hundred responses, written by students at levels ranging from freshmen to seniors, remedial to graduating engineers.

Respondents were asked to give their ages and their race and/or ethnicity. I have included the information about race and ethnicity whenever it was pertinent, so that readers might identify with their experiences and difficulties—and triumphs. I do mention age. Attitudes about the need for mathematics have changed over the years, and older people should know that it is not too late for them to become lovers of math.

Some people merely said something like: "I loved algebra, but hit bottom with geometry." Others wrote a page or two describing incidents in their childhood that had led them to fear math for years afterward. It was heartening to discover that some people, although they had done poorly in elementary or secondary school, had overcome their handicaps to the point where they were taking advanced college mathematics courses or engaged in occupations that required mathematical knowledge.

Mary Jo Cittadino (age 50 at this writing), whom I quoted at the beginning of this chapter, is a success, a person who overcame her fears and avoidance to become a mathematics instructor in the EQUALS program at the Lawrence Hall of Science, University of California at Berkeley. She writes so eloquently about her history and feelings that I use pieces of her autobiography to introduce the first eight chapters of this book.

Here are the words of Verna, who had recently returned to college. She expresses so well the fears and loss of self-esteem that many people suffer as a result of math anxiety:

> I "bypassed" math for as long as possible until I discovered I needed remedial math. . . . I cannot accurately describe my fears. I became so terrified that I was on the verge of withdrawing from college. I felt I would *never* succeed. The fear caused mental paralysis. The more I attempted to overcome the fear, the more "blocked" I became. The more "blocked," the more futile the attempt. I dreaded the exams. I became hysterical and then self-loathing set in; "I was a failure, and I failed because I was stupid." That describes the general cycle. I disliked math because I had such a fear of it.

> I struggled through remedial several times until finally, I passed after a minisession offered in the summer. I began to like math when I began to succeed (passing the exams). My fears have not been totally alleviated . . . , but I don't think I will return to my initial paralyzing fear. I'm at present about to complete a course in statistics (psychology); it's a bit rough going, but I'm not afraid to try.

Deena (age 18), just entering college, has not yet seen a way out. She wrote: "Fear in elementary school, fear in high school, still nervous. If I knew how to overcome it, I wouldn't have the feeling of fear."

Although Laura (age 32) had used mathematics in her work, she did not escape the fear of college math. As a commercial underwriter in insurance for thirteen years before returning to college, she had used what she considered "concrete sensible math." When she transferred from a junior to a four-year college, she was required to take calculus. "I almost died! I got pains in my side." Before a test she would get headaches that intensified as the exam day approached. The night before the test she would throw up. The professor, a "difficult, unkind taskmaster," ignored her pleas for help, and the A's she had earned in her earlier math courses did not increase her self-confidence. "I felt I had received them by good luck, even though I really worked hard! I felt I did NOT deserve them somehow." She could not explain the origin of her fear, but suspects she had internalized the stereotype that girls couldn't do math. A "Mind Over Math" workshop, which dealt with her anxiety from the psychological angle, helped her to get her fears out into the open, and to realize that she is one of many with similar problems. One aim of this book is to give you the equivalent of such a workshop: a no-fault approach to overcoming your fear of math.

Most of the math autobiographies I quote are from women. Like the teacher in London, I found that far fewer men than women were willing to analyze and put down on paper their negative feelings about mathematics. Men were more apt to describe how they had "conquered their weaknesses." Les and Jay, both juniors at a historically black university, are typical. Les recalled: "In high school I performed very poorly in math. However, I continued to take math courses. Having a desire to 'conquer my weakness,' upon graduating from high school, I chose to major in electrical engineering. Although I performed poorly in high school, I have excelled in math thus far in college. I did not overcome my fear of math; however, I used my fear of math to motivate me to study."

Jay compares doing math with athletic practice. "I overcame my fear of math when I realized it was something you had to work at and practice a lot, just as though you were preparing for a ball game."

Marvin, a premed college senior at the same institution, had neglected mathematics in favor of other subjects. However, in college he realized that he needed to bolster his mathematical ability. He followed his own advice: "Put fear aside and do more math. Realize it requires hard work just like anything else. . . . I attacked math and inundated myself with math-related material. I then realized that it just required self-confidence and a willingness to conquer this beast called mathematics."

Mike (age 21) is a junior in college. "[In elementary school] I didn't like the idea of doing hard math, math that required a lot of thinking. . . . [High school] is when the thought of doing math made me *sick* the most, because my math courses grew harder. By the twelfth grade I became interested in math and working with numbers. Now I am a *Math Major;* that should answer your question."

Phil (age 28) had a mixture of good and bad experiences in elementary and secondary school. Ten years later he decided to enter college. Now in a calculus class, he has chosen to remain a mediocre student.

> One problem I have with math is that with any other subject I can guarantee myself an A if I just attend regularly and put in relatively minimal effort. Math requires me to put in more effort for less gain than any other subject. . . . I don't think anything can replace hard work and continued studying. I could get an A out of any math class that I've ever been in but sometimes I just avoid the work in favor of last-minute cramming before a test. It is nice to be able to stay on top of the subject until test time. The feeling of confidence you have when you look at the exam and it doesn't look like it was written in a foreign language is great, although for me, rare.

Phil reminds me of the smoker who says he can quit any time he feels like it, but never does—until it has affected his health and it's too late.

Alan (age 29) was one of the few males who was willing to discuss his math avoidance, a problem so severe as to affect his teenage social life.

> It seems, as I recall, that I've always avoided any situations dealing with mathematical problems. In school I would just "get by" with a passing grade. . . . In real life situations again I tended to avoid math. I remember not being able to play pool and avoided it rather than practice as a teenager. . . . Now I'm attempting and succeeding at learning the one subject I've had the most fear of. It's just a matter of sitting down and not avoiding. I do my homework assignments and regard them as a challenge rather than as a chore. I find myself having to relearn basic skills. . . . Overcoming math problems has been a humbling experience. I'll just keep working at it until I get it right.

Len (age 26) felt that the worst thing that happened to him in math was when he was introduced to fractions. In college he repeated the "basic skills" course several times before he passed. According to his instructor, Len refused to take responsibility for his learning and blamed others for his failure. She believes that he finally matured when he participated in a gay demonstration. Len wrote that now "math is coming along quite smoothly, no real problems. . . . To think back, it's hard to remember why the fear was so strong!" Perhaps Len's main problem was not really math.

The computer has brought new fears to many people. Fear of the computer can be even more devastating than math phobia. Elementary school kids (usually male) handle these complex machines with the greatest self-confidence, while their mothers and their teachers (mostly female) are afraid to touch them. At a New York computer show, according to a story in the *New York Times* (2 April 1984), a woman confessed: "I'm walking around here in a fog. I'm so far behind that I'd have to overcome my computer phobia before I could even sign up for a course on it." Nearby her fourteen-year-old son was discussing technical details of his own work with computers. "I've tried to teach her"—he nodded toward his mother. "She's smart enough. She's just scared."

Many of my respondents were able to trace their fears to specific incidents in elementary school. Often the problem begins with the failure to understand some concepts. A student is out sick, or transfers to another school, or just doesn't get it when the topic is taught. The result—fear of all math.

Now an elementary school teacher, Charlotte was so bright that she was skipped from third to fifth grade. Unfortunately she missed learning long division. "I developed a good case of math phobia and avoidance behavior as a result. I decided that I was a failure at math and that I didn't have a good brain for math." This attitude stayed with her throughout high school and college. Finally she was able to overcome her fears in some stimulating graduate mathematics courses for teachers.

Betty traces her fears to an extended hospitalization in first grade. All through elementary and high school she disliked math and avoided it as much as possible. Enrolled in college at the age of 28, she wrote: "I find myself avoiding taking the math entrance exam for math placement. I've learned to refer to the test as the 'plague.'" So paralyzed was she that she had made no effort to investigate her institution's remedial program. Some time later, Betty enlisted the aid of a friend and, to her surprise and delight, passed the dreaded exam.

The mother of a fourteen-year-old son, Keisha (age 29) is taking a remedial math course at a community college. An asthma sufferer, she missed school frequently as a child and fell behind. When a math test was announced, she just skipped classes the day of the test. In seventh grade she became pregnant and transferred to a continuation school, but at no point did she ever receive the help she needed to catch up. Now she is ready to start life and math afresh.

Alva is an inspiring elementary school teacher in New York City. Her sixth-grade students write plays on current subjects, which they perform for the whole school. At a City Council hearing on South Africa, Alva's class dramatized the horrors of apartheid, contributing to the passage of an anti-apartheid law. But Alva is so terrified of mathematics that she refuses to teach the subject.

Numbers and Numerals

Numbers are everywhere. It would be difficult to imagine our world without numbers. It was the invention in ancient India of base-ten positional notation along with the ten digits—0, 1, 2, 3, 4, 5, 6, 7, 8, 9—that made possible our facility with numbers. By arranging the digits in specific ways, and including a decimal point (or a comma in some cultures), we can represent any number. Each position in this arrangement has a value that is ten times the value of the position to the right of it.

Centuries elapsed before this efficient notation, spread by Arab and north African scholars and traders, was accepted in Europe as a replacement for Roman numerals. Eventually computation with these Indo-Arabic numerals won out over the old counting boards. In those days only the most learned people could cope with computational methods that we now expect elementary school children to master!

Example: The number 306.04 is written in a modern form of Indo-Arabic notation. This is a shorthand way to express:

$(3 \times 100) + (0 \times 10) + (6 \times 1) + (0 \times 0.1) + (4 \times 0.01)$

Here is another way to write the expression, using exponents to show powers of ten:

$(3 \times 10^2) + (0 \times 10^1) + (6 \times 10^0) + (0 \times 10^{-1}) + (4 \times 10^{-2})$

Now imagine how you might multiply 306.04 by 27.9 using Roman numerals!

My fear started in fifth grade. The teacher, a nun, would send me to the board to do math exercises. Meanwhile she would be looking at her watch and timing me: "Every second that you take comes off the boys' gym period." The boys would call out to me to hurry, and my friend would try to give me the answers without letting the teacher find out, but that didn't help, because I couldn't make out what she was whispering. I was the

> only black child in the class, and that may have been the reason for this treatment. I never told my parents how I was being treated.
>
> In college I had to take just one math course—algebra. I took it in summer school to get it over with, and there wasn't time to learn much. I was not required to take any math courses for teachers.
>
> I have taken some math workshops since, but I can't get over my fears. My husband says that math is just logic and common sense—"It's simple!" But not for me!
>
> This year I have the brightest sixth grade class. I don't want to mislead them, so I have arranged for my colleague to teach my class math several times a week, while I take over her class for another subject. That works out well.

Unfortunately, this exchange arrangement did not last beyond one academic year, and Alva had to find other ways to solve her problem.

For Yvonne, too, the fear started in fifth grade with a teacher who made math "hateful." Math continued to be hateful for her through graduation from a community college and remedial math at a four-year college. But she persists: "I intend to conquer this fear and move onwards to trigonometry."

I met Isadoro (age 50) at a local school. He had arrived from Cuba at the age of 6, and entered school knowing no English. He can still recall the misery of his first few years in elementary school; too scared to talk, he was rated poor in both reading and math. When his fourth-grade teacher slapped him for not doing his work, he was convinced that he was hopelessly stupid. His college experience consisted of a few courses here and a few courses there. After failing algebra several times, he enrolled in a class that dealt with his math fears and discovered that he could handle math concepts quite well. Finally, at the age of 38, he earned his bachelor's degree. Now Isadoro teaches math and science to middle-grade bilingual students and is quite confident about his ability to put across the important concepts. He impressed me as a very empathetic teacher.

As a graduate student in wildlife biology at a California university, Diana (age 27) finds now that she needs the mathematics she had been avoiding all her life. The degree of her avoidance depended mainly upon the attitude of her instructors after an unsympathetic third-grade teacher had triggered her fears. Now she is coping with the statistics and calculus required for her career. She attributes her current success to the sympathetic approach of the instructors in her "Math Anxiety" class and calculus workshop. Much to her surprise, she discovered that she could do math when she tried. She realizes, as do many others, that her feeling of inadequacy was due to avoidance, rather than to incompetence.

Many institutions now offer summer "bridge" programs to help students to make the transition from high school to college and to enable them to make up some of their deficiencies. Tomas (age 17) was taking pre-algebra in a California college bridge program when he wrote his math autobiography. Typically, his fear of math had started in fourth grade, and by junior high school he "hated math with a passion." His final paragraph expresses his conflicts: "I personally feel I can't overcome my fear or dislike for math for the reasons that math is just getting harder and more confusing. I'm afraid of it. It doesn't bother me, though. I'll get by somehow."

Is this a case of male bravado, an unwillingness to face the issues squarely and admit that he *is* bothered? One would hope that Tomas will find in college the kind of sympathetic instructors who have helped other victims of fear of math to overcome their dislike and feelings of inadequacy.

Are elementary school teachers really as cruel and lacking in understanding as portrayed in some of these case studies? Probably most of these teachers would be shocked to learn how their former students viewed them, when they were just trying to do their job! Faced with the task of inculcating the prescribed mathematical skills, and called to account when their charges did not measure up to certain standards, they did their best. Many elementary school teachers are like Alva, aware of their fears and their poor mathematical background, but unable to arrange her intelligent solution to the problem.

Contrast Alva's experience with that of June (age 41), now in the third year of an undergraduate program for future elementary school teachers.

> I struggled with algebra and failed geometry. In 10th grade I had one of the most frightening teachers I've ever

> had. When I went to him after school seeking help he yelled at me and threw the answer book at me. That was my last high school math experience. I felt that I was incredibly dumb. . . . During my junior year [in college] I took Elementary Math Methods . . . and have learned how to make math (learning) a wonderful experience for children.

Parents, too, can steer their children in the wrong direction. They may have been influenced by the stereotype that girls are not capable of doing math, or they may believe that women's careers do not require a knowledge of mathematics. This was true in the case of Lucille (age 27). When she did poorly in junior high school math, her father consoled her by saying: "It's OK if you don't take math. I never liked math. You don't really need it." This turned out to be the wrong advice. As a photography instructor she definitely does need math. She feels inadequate when she is called upon to explain the technical aspects of photographic processes, and is thinking about going to a math clinic for help.

How many times have my students told me: "I really did understand that topic, but I just froze on the test! I couldn't remember a thing." Yes, taking a test can be very stressful, and particularly so when the material is only poorly understood, or merely memorized rather than understood. Terry (age 17), a first-year college student, wrote: "I feel that the reason why I didn't like math is because I didn't do well because I just plain didn't understand it. Why? I really don't know why. I guess because I didn't take an interest in it because I didn't understand. It was a vicious circle that I went through with math. And when I did understand, I would freeze on a test and almost forget everything. I don't think I will ever be good or even like math but I have a positive attitude that I will learn it."

For Tomi (age 21) test taking is a painful experience. Her previous calculus teacher had "let us retake the tests and gave take-home exams. By this time I felt I really knew the material." But the following semester she was doing poorly. "I don't think I'm stupid but due to lack of time on going over certain subjects the material just isn't sinking in like it did last term. I wish we could rework our tests. It would make such a difference in whether or not the material gets learned and stays with a person. I just seem to blank out on the first part of the test. Midway through I start to loosen up and get rolling. I've always done better when I've had more time to take a test."

Alberta (age 20) was disheartened when she was placed in a developmental (remedial) math class in college, since she believed her real problem was fear of taking tests. She did well enough to enroll in college algebra the following semester. Unfortunately the instructor was not very helpful. "I signed up for tutoring because we began word problems—'The Big Chill.' The tutor that I had made me feel dumb, so I never returned. She expected me to know certain things but I just got discouraged, so I ended up failing the course. I tried to take the class again, but the fear of failing the class caused me to withdraw." Alberta is insightful about the way mathematics is taught. She had started to feel uncomfortable with math during her high school years, and with good reason. "I don't think that math should be just given to you from the a book," she wrote. "Math should be related to other subjects and to life situations." I agree with her wholeheartedly!

The message in many of the math autobiographies is that everyone can learn math, in spite of a poor start in the early grades, or failing grades of high school, or poor teachers along the way.

Tammy (age 18), in her first year at college, wrote: "I love math. I didn't learn it in high school. I learned it later in college. I think it's a great subject."

Vera (age 19) had a similar experience. "By my second quarter in high school, I had started to develop a dislike of math. During my senior year I had a teacher with a negative attitude who helped to reinforce my own negative attitude toward math. When I came to college, I had a very supportive class and teacher. Both the teacher and my classmates helped me to develop a more positive outlook toward math. Now, math isn't so bad."

Some people are not as fortunate as Vera and Tammy, but they stick to it and are determined to overcome their difficulties. The transition from schooling in another country to college in the United States can be a painful experience. Pamela (age 44) is a good example. Coming from Jamaica, with a limited background in mathematics, she said: "Most times it seems like a foreign language to me. But I have a great desire to master this subject."

Deepak (age 23), a recent arrival from Bahrain, wrote: "Calculus was a terrible experience for me since there were lots of things I did not have prerequisites for. My fear was overcome when I passed Calculus I. It required lots of concentration and practice."

Rahmon (age 19), also a recent arrival to the United States, had been unable to cope with the unpleasantly strict high school mathematics teacher in his Asian homeland. Although he did fairly well in his first college calculus class, he was jealous of his friends who earned A's, "but I can't do anything. That becomes my nightmare. Sometimes I talk to myself: 'It is very peaceful if there is no math in this world.' I know that is very silly, but that's all what I feel." Now he is beginning to overcome these feelings of inadequacy. "I'm trying to learn to like math by studying it very relaxed. I'm not trying to memorize all problems, because math can't be understood by memorizing, but by understanding and trying to do the problems; and it really works. I got a good score for my first test. It was an amazing thing in my life."

Age is not a drawback, and may even be an advantage, as Rose will testify. The mother of twelve children, she decided at the age of fifty-six to go to college and take basic math. Although she finds the work hard, and her children tease her, she is persevering. "I was too embarrassed to ask questions when I was young. Age does make a difference."

Many galleries and institutions display the fine bronze sculptures by Judith Weller (age 54). Most notable is the *Garment Worker at the Sewing Machine,* set on the pavement in the heart of the New York City garment district. Judith recalls her high school math teacher in Israel as a tyrant whom the students loathed. Her father, a tailor and the model for her sculpture, relied on Judith to deliver the garments to his customers, resulting in her occasional lateness to math class. The teacher shamed her before the class and maligned her to other teachers. The effect on her work in mathematics was disastrous.

About two years before our interview, Judith thought she might go for a graduate degree in architecture, and she enrolled, with trepidation, in an algebra course. "I loved it! I finally saw what I had missed. This teacher was very sweet and gentle. He respected the students. He respected all their questions, although some students had been away from school for many years. I was so surprised that I enjoyed the course. At best I thought I would tolerate it." Judith took the Graduate Record Exam and did well in the mathematics section.[7]

Florence was one of the oldest respondents to my call for math autobiographies. At the age of fifty-nine she was about to receive her Bachelor of Arts degree. She recalled her anxiety about math in elementary school, and her experience with high school math was even more discouraging. More than thirty years later she had the opportunity to attend college:

> I entered it with no reservations, psyched myself up when it came to take the remedial math course I needed to enter into other courses required for a B.A. degree. At this time I felt very good for I like challenges and math would be mine once more.
>
> I had to repeat the remedial several times. It hurt, but every time I left I knew more than when I had started. Slowly but surely it came to me. Math has been a problem to me, but I know that if I try and try I always learn some of it. Since I am a working person, married and attending college at night, I know that I do the best that I can, and have to accept the outcome.

A common thread runs through most of the autobiographies. These victims feel powerless, out of control, lacking in self-esteem. Many people manage to conceal their negative feelings about math until they are confronted with a situation that forces them to "come out of the closet." Frequently the new element in their lives is enrollment in college, whether after a lengthy absence from academic life or immediately after high school graduation. Having avoided math successfully, perhaps for years, they suddenly find themselves facing a mathematics qualifying test or a required statistics course. Consequently colleges must not only offer the remediation that these people should have received much earlier, but must also help the victims of fear of math to overcome their anxieties about the subject.

One might suppose that students at selective Ivy League colleges would be exempt from such difficulties. Not at all! Professor Deborah Hughes Hallett initiated a remediation course at Harvard University for students "who have trouble with decimals, and who find l.c.d.'s [lowest common denominators] impossibly mysterious, and who drop minus signs and parentheses as though they were going out of style, and who freeze at the very thought of a word problem." Although these students have taken algebra and other high school math courses, they passed by memorizing rather than by understanding the content. But, as Hallett asserts, "People forget what they've memorized as soon as possible." Furthermore, "It contributes greatly to one's fear of a subject to know that you don't understand what you're doing."

Hallett and her associates find that their attitude to the students matters more than their teaching methodology. "They learn far more from our faith that they can and will learn mathematics than from our most lucid explanations or most brilliant innovations. . . . I think you can always learn math—it may take time, work, and energy, but it always can be done.[8]

You might want to counter Hallett's confident assertion by saying: "It may be true that all Harvard students can learn math eventually, but that doesn't go for an ordinary person like me."

That's not true. Everyone really *can* learn math. I had the pleasure of visiting classes in Project Bridge at Laney College in Oakland, California. This program is designed to prepare people with minimal skills (fifth-grade level) for regular college work or vocational training. The students, many of them high school dropouts, range in age from the late teens to the sixties, although a majority are African-American and Latina women in their twenties and thirties. Students in the program are active participants in their own learning and they really care that their fellow students learn as well.

In one class the students were learning about the lowest common denominator, just as the Harvard students had done. First the instructor worked some examples on the chalkboard, while the class participated with suggestions and questions. Then a young woman volunteered to do a problem at the board. As she proceeded, it became obvious that her motive in volunteering was not to show that she knew the procedure perfectly, but rather to work through a difficult situation with the support of the other students. When she finally arrived at the solution, she beamed with delight and took a bow, while the class cheered and applauded. Such warmth and appreciation would draw out the most reticent person!

School math as you have probably experienced it is responsible for an unrealistic view of what math is all about. Some common misconceptions are:

- Math is mainly arithmetic, working with numbers. If you are not good in arithmetic—for example, you haven't memorized the multiplication tables—you can't learn "higher level" math, like algebra and calculus.
- Math involves a lot of memorization of facts, rules, formulas, and procedures.
- You must follow the procedures set down by the teacher and the textbook.
- Math must be done fast. If you can't solve a problem in a few minutes, you might as well give up.
- Every problem has just one right answer, and it must be exact.
- You must never count on your fingers or use hands-on materials to help you solve a problem.
- You must work on math alone. Working with other people is cheating.
- You must keep at it until you have solved the problem.
- Math is hard. Only a genius or a "math-brain" can understand it.
- Math language is unrelated to ordinary everyday language.
- Math is rigid, uncreative, cut-and-dried, complete. It doesn't involve imagination, discovery, invention. There is nothing new in math.
- Math is exact, logical, and certain. Intuition doesn't enter into it.
- Math is abstract. It is unrelated to history or culture.
- Math is value-free. It's the same for everyone all over the world.

This book will offer counterexamples to each of these misconceptions. You can find many more in the books and articles listed in the reference notes and the bibliography.

With this book I hope to achieve the following goals:

First, to give you reasons for overcoming your phobias about mathematics. Why bother learning math? Why not let well enough alone? Who needs math, anyhow? The answer is: Everybody needs math! Why? That is the theme of chapter 2.

Manipulating Math

The more ways you deal with a concept, the better your understanding. Working with concrete materials, called manipulatives, can often clarify confusing concepts. Manipulatives, such as *Base Ten Materials,* also afford the opportunity to experiment and discover new ideas.

Base Ten Materials include:

- small squares, representing *ones,* and often called "units";
- Strips of ten small squares, representing *tens,* and often called "longs";
- Large squares measuring ten units along the edge, representing *hundreds,* and often called "flats."

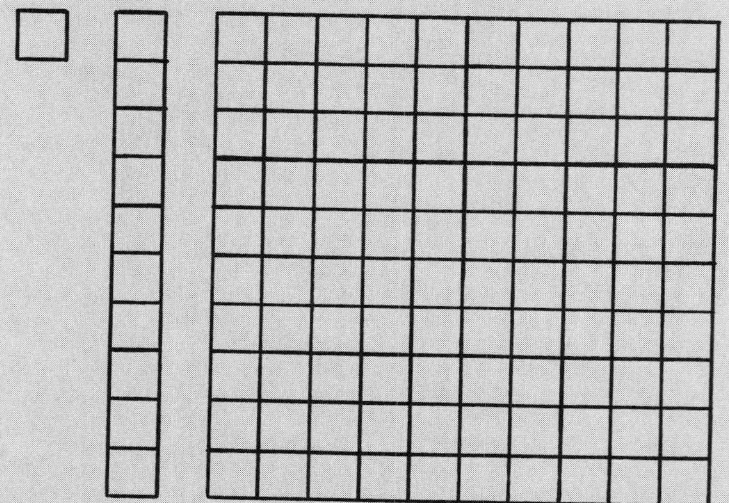

Base Ten materials may be purchased commercially or made at home of stiff paper or plastic sheets. They are useful for all types of number work because they embody our base-ten numeration system.

(*Continued*)

Manipulating Math (*Continued*)

Example: Find the product of 12 and 13.

Procedure: Construct a rectangle that measures 12 units on one side and 13 units on the adjoining side, using as few pieces as possible. Start with the largest piece, the flat. Then add as many longs as you can fit into the rectangle. Fill in the remaining spaces with units.

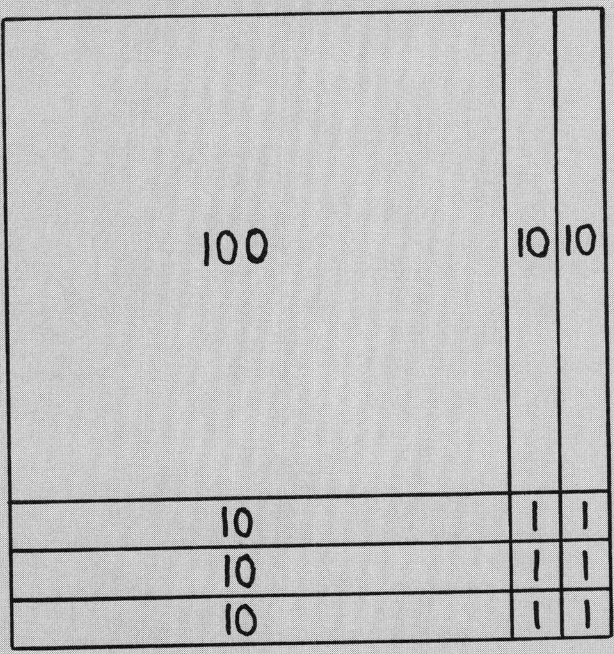

Now figure out the total value of the pieces in the rectangle:

1 flat	= 100
(2 + 3) longs = 5 × 10 =	50
(2 × 3) units = 6 × 1 =	6
Sum	156

Compare this method with the pencil-and-paper procedure, step by step. You have also found that the area of a 12 × 13 rectangle is 156 square units.

Second, to help you see through the myths about who can and who cannot learn math. But—you might be thinking—what if I just don't have what it takes? Don't you have to be a genius to do math? Some people say that girls are not capable, that blacks are not capable, that Latinos are not capable of doing math. Don't believe such myths. The false theories that give rise to these myths will be treated in chapter 3.

Third, to understand how parents, friends, teachers, and the larger society may have been guilty, perhaps not intentionally, of steering you in the wrong direction, away from the kind of intellectual work that can make your life richer and more fulfilling. Chapter 4 will deal with these issues.

Does the school system really give everyone an equal opportunity, regardless of gender, race, language spoken at home, or social class? Of course not, and that disparity is a source of math fear and avoidance. Math classes are conducted differently for different groups, starting from the earliest grades. Moreover, school math, for the most part, is not real math. Dr. Hassler Whitney, one of the foremost mathematicians of the twentieth century, would say to mathephobes: "You never had a chance to see or do real math, which is easy and fun." Every society has developed mathematical ideas and practices, yet these ideas and practices are generally excluded from the school curriculum. I will discuss these issues in chapters 5 and 6.

Chapter 7 is designed to help you to help yourself—strategies to overcome fear and avoidance of math. You probably know more mathematics than you think, and you must give yourself credit for this knowledge. You will learn some techniques to reduce math anxiety and how to accommodate your unique learning style to the material to be mastered.

As a parent, a teacher, an employer, or a friend to young people, your ability to overcome your own fears and improve your skills will also help you impart to children a love of mathematics and self-confidence in their ability to do math. Chapter 8 is intended to help you do that.

Lastly, chapter 9 is a call for a more humane mathematics, the kind of math that will serve all people.

Interspersed throughout the book are real-world problems for you to think about as you read, or to come back to later. Does it matter whether I get a raise or a bonus? Can all children be "above average" achievers? How likely is it that a single woman over age 30 will eventually marry? Are men or women doing better in the campaign against smoking? What percentage of the federal budget is actually used for military purposes? The examples, covering a wide range of topics and of levels of difficulty, should help you to understand the math that you encounter every day as a citizen of this planet.

Cornell University professor David Henderson has these encouraging words to light your way:

> I believe that mathematics can be a part of every person's understanding and can have an important role in the liberation of human beings. I define liberation as the removal of all barriers to a person's creativity. . . . Every person who needs some part of mathematics in order to understand some aspect of their experience can grasp that part of mathematics in a very short time. All that is needed is confidence in their thinking and in their perception. This assumption applies, I believe, regardless of the person's mathematical background. . . . I am certain that as women, and members of the working class and other cultures participate more and more in the established mathematics, our societal conception of mathematics will change and our ways of perceiving our universe will expand. This will be liberating to us all.[9]

CHAPTER TWO

Who Needs Math? Everybody!

> My first job was as a junior secretary at Wayne State University in Detroit, Michigan. They told me they almost didn't hire me because I'd scored so poorly on the math portion of the pre-employment test. My verbal scores were so high they decided to risk it.
>
> After one semester on the job . . . I decided to apply for admission as a night student. Because I'd had no math or science in high school, I was asked to take an admissions test. Because of my scores (high verbal/low mathematics), they admitted me on probation. I'd have to maintain a C or better average or lose the privilege of attending.
>
> [Many years later] something happened at the checkout counter of the grocery store. The bill seemed too high, but I was afraid to question for fear of being wrong and, therefore, humiliated. When I got home, I checked each item against the cash receipt and discovered I'd been charged $12 for three quarts of milk. I had to make a special trip back to the grocery store. Enough was enough. I had to change.
>
> —*Mary Jo Cittadino, Mathematics Educator, EQUALS, University of California at Berkeley*

Opportunity is knocking at the door. Future job prospects look brighter than in the past for women and people of color, the very groups that have until recently been excluded from many well-paid careers. The forecasters tell us that by the year 2000, about 85 percent of the new entrants to the work force will be women and members of "minority" groups. In a speech calling on business leaders to contribute to the training of young people of color, New York Governor Mario Cuomo said, in what seems to be a contradiction in terms: "The majority of the work force in this country in the twenty-first century is going to be minority."[1]

At the same time there will be a greater demand for workers with a high level of skills—scientists and engineers, teachers and managers. No longer will a high school diploma serve as entry to most jobs. Advances in technology will require well-educated workers, people who can adapt to new conditions. It is estimated that, with the rapid changes in the nature of employment, young people today can look forward to as many as five different careers in their lifetimes.

Past Discrimination

Until recent times, only an elite class of people, mainly white middle- and upper-class males, required more than a minimal mathematical background for their jobs and careers. Barriers of race, ethnicity, and gender excluded many qualified people from math-related professions, as I know from my own experience. I had earned a master's degree in actuarial science, the mathematics of insurance, and had passed several of the actuarial exams while still a student. I soon discovered that the field was completely closed to women and that there was exactly one employed Jewish actuary in the entire United States. Since I was both female and a Jew, my education and achievements in actuarial science seemed to have been wasted. After the Second World War this type of discrimination lessened, and I began to receive job offers from various parts of the country. By that time I had a family and was no longer in a position to accept these attractive offers.

In 1949 Evelyn Boyd Granville was one of the first two African-American women to earn a doctorate in mathematics. After graduation from Smith College with highest honors, she was awarded a fellowship to Yale University, where she studied with a renowned mathematician. Nevertheless, her application to teach in a New York college was greeted with laughter. The students at Fisk University, a historically black institution, were the fortunate recipients of her wisdom. Later she spent sixteen years in applied mathematics for the government and private industry, where she used her supurb talents to contribute to the "space race."[2]

Discrimination in higher education was blatant. My husband was teaching electronics at a two-year college in the 1950s. As part of the job, the department faculty members were required to interview applicants who planned to major in electronics. The chairman had directed the instructors to mark the application forms of all black applicants, with a view to barring them from admission. "They won't get jobs in the field, so why disappoint them?" was his lame excuse. Other members of the department refused, citing both the illegality and the immorality of such an action, and forced the chairman to withdraw his racist directive. No doubt, many people were reluctant even to apply for such programs for fear that they would not find employment afterward.

The civil rights and women's movements of the sixties and seventies were instrumental in securing passage of legislation to lessen the inequities in employment for women and people of color. Affirmative action programs have resulted in important gains for these groups. In the period from 1973 to 1980, for example, female employment in the work forces of government contractors grew at a rate seven times that of nongovernment contractors, who were not subject to regulation, while the rate of increase for minorities was almost double. Moreover, many more low-skilled females and minorities who worked for government contractors moved up to high-skilled and white-collar jobs, as compared with workers for nongovernment contractors.[3] In the 1980s, however, many of the programs were phased out or weakened, due to cuts in federal funding and to lack of enforcement of equity laws. Today, in the 1990s, the situation is different. The employment of women and people of color is no longer viewed merely as an issue of fairness and social responsibility. Now it is a matter of national survival. Our economy needs them!

The Need for Mathematics in Today's World

The economy of the United States has changed drastically within the past few decades. There has been enormous growth in communications and finance, while manufacturing jobs have gone abroad or have been restructured to require higher-level skills on the part of the workers. With advances in technology occurring frequently, people will have to learn to work smarter, rather than faster, than in the past. Increasingly the jobs in tomorrow's economy will require a knowledge of mathematics. New technologies will call for the ability to apply mathematics and science in practical ways, and rapid changes will demand

Women and Minorities Use Statistics to Win Their Case

Knowing the numbers helped women and members of minority groups to win a bias suit against a Chicago bank. In January 1989 the bank was ordered to pay $14 million to settle a case charging that it had discriminated against women and minority group members in its hiring and promotion policies. Statistical evidence was crucial in winning the largest monetary settlement in a bias case to date. The case had been initiated in 1974 by the organization Women Employed. The U.S. Labor Department had brought suit on the basis of a federal antidiscrimination order. (*New York Times*, 11 January 1989, A1.)

that workers learn new skills throughout their lives. In other words, lifelong learning will be the pattern of the future.

In the old days, blue-collar workers stood at an assembly line or a machine and performed the same few operations over and over again. Little formal education was required for this kind of job. With strong unions to back them up, these workers could earn adequate salaries.

But new blue-collar jobs are becoming rare. A recent test for sanitation workers in New York City brought 101,000 applications for the expected 2,000 openings over a four-year period.[4] Although a high school diploma is not required, the test included such mathematical skills as "recognizing how an object will look when it is moved around or when its parts are moved or rearranged; applying general rules to specific problems to arrive at a logical answer; adding, subtracting, multiplying and dividing numbers."[5]

Many blue-collar and clerical workers are now being displaced by high-tech machines. According to Ronald E. Kutscher, associate commissioner at the Bureau of Labor Statistics, "People who have less than high school education are in a difficult position whether they're blue collar or white collar. They have higher

A Bonus or a Raise?

Which is better for a working person, a bonus or a raise? Many businesses have been giving their workers bonuses based on profits, rather than a percentage raise in wages. In 1988 Chrysler workers felt compelled to accept a 3 percent annual bonus, while the expiring contract had provided for a 3 percent annual wage increase (*New York Times*, 10 May 1988, A1). Even when the bonus represents a larger percentage than the wage increase, workers usually lose out in the end. Why is this so?

Let's work it out both ways for a period of three years. Assume that the worker received $28,000 in wages in 1988.

Bonus method: For three years the worker receives $28,000 annually plus a bonus of 3 percent of $28,000, or $840. For the period 1989 through 1991 he will earn (3 × $28,840), a total of $86,520.

Percentage increase method: In 1989 the worker receives $28,840. In 1990 her increase is 3 percent of $28,840, or $865, making a total of $29,705. In 1991 her raise is 3 percent of $29,705, or $891, for a total of $30,596. The three-year total is $89,141. This sum is $2,621 more than she would have earned with a bonus.

But there is more to the story. Benefits like vacation pay, overtime, severance pay, and some pensions are based on the wage rate for each year. A larger base wage means larger benefits. It all adds up.

It really pays to know your math, provided that you and your fellow workers have the clout to achieve your demands.

unemployment rates and they have more trouble finding a job. . . . The more resources one brings to the job market, the greater one's ability to rebound. Education is one of those resources."[6]

Workers in the new type of "informated" factory are expected to take full responsibility for computerized equipment and robots, to make their own decisions about production and scheduling, and to be able to track problems. Looking at the Japanese success in manufacturing, experts in the United States found that many of the Japanese who work directly with automated machines have advanced college degrees. This is the trend in the factory of the future—fewer low-skilled laborers and more engineers, scientists, and technicians with solid mathematical backgrounds.[7] Many corporations are instituting their own training programs to upgrade their workers' skills.

Of course, not every job requires specific mathematical skills. But it certainly helps to have some facility with math and some confidence in your ability to handle problems as they arise. Entry tests and civil service exams for many jobs, like that for sanitation worker, usually include questions involving mathematics. The employer or agency may justify this practice by claiming that such questions reveal whether the applicant is capable of coping with new situations. More likely, these tests serve as filters to keep out applicants with inadequate educational backgrounds or poor skills in test taking.

The greatest growth industry, say the futurists, will be information processing, and computers will play a major role. Already computer skills are necessary for many types of office work and business management positions—you must know a spreadsheet from a bedsheet!—and workers are expected to be familiar with complex information-processing systems. In fact, if the computers were to break down, operations in the world of commerce, finance, insurance, and communications would come to a standstill.

It is not only business that requires a knowledge of computers. The computer has also invaded the fields of health care, music, theater, fine arts, and many others. At many art schools, for example, students learn to use the computer in their design classes, with additional applications to the areas of sculpture, ceramics, photography, interior design, and printmaking. Not that technology is replacing artists in the creative process. The computer is a "tool, a flexible and pliable tool, not meant to replace the artist with technology."[8]

Eleanor (age 30+) was my instructor in a course in the tie-dyeing of fabrics. An innovative artist, she finds that math, physics, and chemistry are absolutely essential to her work. Luckily she has always enjoyed these subjects and, even more essential, has confidence in her ability. One of her art projects involved setting up a tremendous tent. Without her working knowledge of dynamics and the properties of various materials, the tent might have been blown away by the wind. In order to dye textiles she must understand chemistry and be able to calculate appropriate times and temperatures. Of course, geometry also plays an important role in her art.

Vicki (age 37), struggling with her third semester of calculus for a master's degree in architecture, had taken very little mathematics for her college major in anthropology. "When I decided to change fields and attempt architecture, I knew that I would be taking more math." Up to that point Vicki had felt that math was "irrelevant to my life and interests. I do not feel that way now, and wish that I hadn't been so blind earlier, either to the nature of mathematics or to the real nature of my own interests."

Both Olav (age 32), a professional bicycle racer, and Tony (age 28), a construction worker, are studying mathematics at a community college with the goal of changing their careers. Tony is taking geometry and algebra to regain his mathematical skills in preparation for a degree in engineering. It wasn't until he started working that he found mathematics to be useful. He advises that "real world applications and examples are a must for all math classes."

Herbert, too, needed to change his job, but not for the usual reasons. His wife was unable to become pregnant, and their application to an adoption agency had been turned down because their income was too low. In a later chapter we will come back to Herbert and learn how he was helped to prepare for a new career by overcoming his frustration with learning math.[9]

Changes in the economy and in their financial responsibilities motivate many people to plan new careers. In fact, they are returning to school in such numbers that two out of five college students in 1988 were adults over the age of 25. Many more study outside of schools. According to the National Research Council report *Everybody Counts*: "Almost as many persons study mathematics outside traditional school structures as inside them. . . . Today much mathematics is studied by older adults. Some large businesses actually operate mini-school districts just for the continuing education of employees; many universities and colleges—especially community colleges—attract large

Household Income: Looking behind the Statistics

More women have been entering the work force in recent years. In the late 1970s half of mothers with children under the age of eighteen were working. By 1990 the number had increased to two out of three. Many women with husbands in the labor force find that they, too, must work to keep their families afloat. Out of every ten women with husbands present, four were working in 1970, five in 1980, and six in 1990. How has women's participation affected household income, the income of all persons living in one household?

I did some figuring, with the help of my almanac and my calculator, and listed the numbers in a table:

Household Income in the United States

Year	Current dollars	CPI	Adjusted dollars
1970	$ 9,867	38.8	$25,430
1980	21,023	82.4	25,513
1990	33,956	130.7	25,980

Here is the explanation of the figures in each column:

Current dollars: The average household income for each year in actual dollars.

CPI (Consumer Price Index): The cost of living for that year as compared with the cost of living for the base year 1983. For further discussion of the CPI, see "Inflation: Your Shrinking Dollar" (page 40).

Adjusted dollars: Divide Current Dollars by CPI for each year and multiply the answer by 100. These figures represent real income.

The numbers in the last column are almost identical for each year: 1970, 1980, and 1990. The American family had to run like hell just to stay in one place, even with more wives going to work. To make matters worse, the adjusted income for 1991 was even lower than the 1990 figure, and was less than the income for 1979, according to the *New York Times* (4 September 1992, A1).

numbers of adults both to regular degree programs and to special short courses for professional growth or cultural enrichment."[10]

More women are now entering the work force than ever before, usually out of economic necessity rather than choice. Traditional women's jobs and low wages go together. To better their position, women would do well to take advantage of the opportunities opening up in nontraditional fields, a preparation that often involves knowledge of mathematics and computers.

Both for young people just entering higher education and for adults contemplating a change of career, it often comes as a surprise that they need mathematics. "Why must I take math to be a social worker?"—or a nurse or a psychologist? But these "helping professions" now have specific requirements for courses in algebra, statistics, computer applications, and more. Florence Nightingale, the woman who established nursing as a profession in the mid-nineteenth century, also developed statistical techniques that she used to convince her government to appropriate money to reduce the number of deaths by improving conditions in hospitals.

Mary (age 34) met her nemesis in a required course in statistics. A college honor student, she had planned to become a psychologist, but after twice failing statistics, she changed her major to social work. She had hated and feared high school math and consequently had taken no more than the courses required for an academic diploma. "Math still makes me feel dumb," she wrote in her autobiography. "I won't do our taxes or balance a check book, but now I teach people how to do budgets, and that feels OK." Apparently Mary acquired the skills necessary to handle the mathematics required by her job as director of a funding agency, but not the self-confidence to overcome her old fears and aversions.

Lori (age 46) had studied the usual high school math. She had earned an A in algebra but couldn't see how geometry would relate to her life as a female. She went on to teach English and business subjects. "I had no fear of math except statistics in college at the master's degree level. I had heard it was a difficult course and feared it. I came out OK, but felt I needed a better background at earlier stages in school."

On the other hand, for Esther (age 32) the required college statistics was the first mathematics course that she really enjoyed. Not only was it relevant to her major in sociology, but she felt that she really understood underlying mathematical

Men's and Women's Earnings: Looking behind the Statistics

We know that women earn far less than men, even when they are doing the same job. For a long time, until about 1981, women who worked full time were earning between 57¢ and 60¢ to a man's dollar. The good news is that after 1981 the difference grew smaller, reaching a high of 71¢ to a man's dollar in 1990.

Now comes the bad news. While women's earnings remained constant in the late 1980s, men's income was actually falling and had been falling for some time. Women gained only in relation to men's declining earnings, but

(*Continued*)

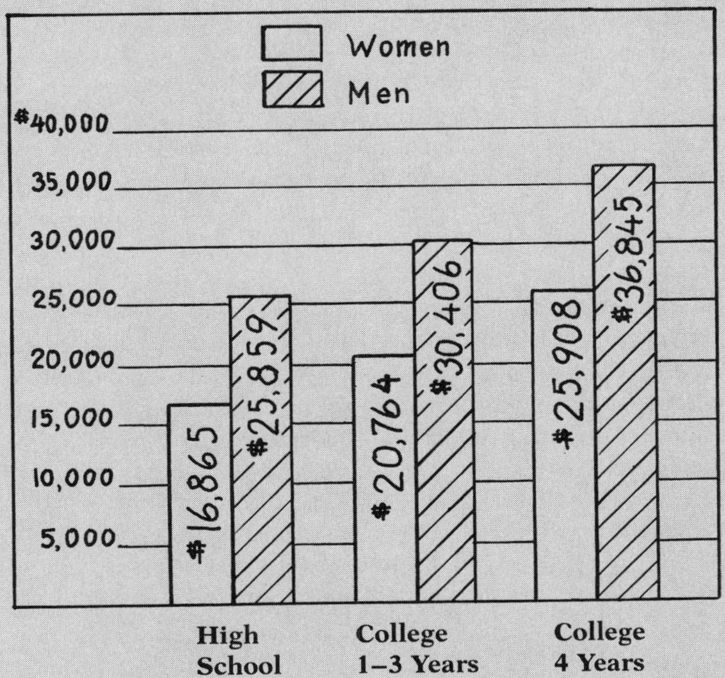

Education Pays Off! Earnings of Year-Round Workers, 1989 (persons 25 years and over)
SOURCE: Department of Commerce, Bureau of the Census.

Men's and Women's Earnings: Looking behind the Statistics (*Continued*)

not in absolute terms. Besides, factors such as the increase in male unemployment were not taken into consideration. Level of education made a great difference in the earnings of both men and women, as the graph shows. (U.S. Department of Commerce, Bureau of the Census, 26 September 1991.)

concepts for the first time. As a hospital administrator, she also makes practical use of her statistical knowledge.

The experience of Janine (age 50) is heartening. When she was growing up in Tennessee, it was taken for granted that, as a woman, she needed no secondary-level math. Now a widow, she is enrolled in an Early Childhood Education program at a community college. "I raised four children and helped with three grandchildren. I might just as well use that experience in my career." The refresher math courses she has taken have helped her to feel independent. "I can handle my own bankbook!" she exclaimed with pride.

Ginny (age 21) was about to enter medical school when she wrote in her autobiography: "Math was not a subject that I enjoyed; rather, I realized that it was a necessity in my education. Additionally, math is challenging, and I enjoy taking on difficult tasks. . . . I am looking forward to taking computer science courses to keep me in touch with the changing mathematical society." And, she might have added, with the growth in computer applications in the field of medicine.

A recent arrival from Taiwan, studying calculus as a first-year college student, Blossom (age 20) admits that she really doesn't like math, but "I still have to do it because I have to pass the exams and I need the grade." Isn't that the motivation for many students?

Not so fortunate is Moni (age 18). She felt that she did understand high school math, yet she could barely pass a math test. "So at present I am in the pharmacy curriculum and realize that math and chemistry, or courses requiring math, are more than required

How to Tell the Truth with Statistics

"There are three kinds of lies: lies, damned lies, and statistics" is an oft-quoted quip credited to Disraeli, the nineteenth-century British statesman. On the other hand, his compatriot H. G. Wells said: "Statistical thinking will one day be as necessary for efficient citizenship as the ability to read and write." These two quotations appear on the first page of Darrell Huff's popular book, *How to Lie with Statistics.*

To these I'll add a third statement, spoken by a member of the United States Congress. To celebrate the publication of *The American Woman 1987–1988,* a study commissioned by the Congressional Caucus for Women's Issues, Republican Congresswoman Claudine Schneider declared: "Statistics have to be one of the tools we use to change the system. Clearly this book provides us with the facts, the figures, the statistics, absolutely everything we need." (Barbara Gamarekian, "Status of Women Rises but Pay Lags, Study Finds," *New York Times,* 22 July 1987, A23.)

of me. . . . But all I can do at present is to work hard, again, and try to continue to excel in this troublesome course of my life."

Whether or not the major field of study involves mathematics, most colleges require that students pass a math proficiency examination, a dreaded ordeal for many liberal arts students. Ken (age 37) is obviously resentful at having to take such a test. He had already attended college for six years, but was just confronting the college math requirement. "I don't feel I have any anxiety toward math. I have just found it difficult and not relevant to my life. I can't think of any instance in my life when I have needed more than I already know." I suspect that Ken is expressing a typically male "I don't need it" attitude to cover up his lack of confidence in his ability to do math.

Diana (age 35) has a more optimistic attitude. She had taken only one math course in college, the easiest one she could find. Fifteen years later she returned to school for a degree in journalism. To her dismay, she had to take a proficiency exam in math.

Infant Mortality: What's behind the Numbers?

How well does our country take care of its people? One indicator of a society's general health is the infant mortality rate. This rate states the number of infants who die before the age of one year for every thousand live births.

The infant mortality rate has been decreasing steadily. By 1988 it was just under ten deaths for every thousand live births. Should we be proud of this record? Let's look more closely at the numbers.

While the overall rate has been falling, the mortality rate for black babies has not kept pace. In fact, the gap between whites and blacks has been widening. In 1987 black babies died at the rate of 17.9 per thousand live births, more than twice the white rate of 8.6. One factor is the failure of many black women to receive prenatal care in the first three months of pregnancy, generally because they have no health coverage.

When compared with other countries, the United States is not doing so well. This rich country ranked behind twenty-one other nations in 1988 in the rate at which infants die. ("Health Data Show Wide Gap between Whites and Blacks," *New York Times*, 23 March 1990, A17.)

Although she was still frightened, she was determined to do well, since "at this point in my life I am no longer content with math ignorance."

Using mathematics on the job or as preparation for a career is only one aspect of the influence of mathematics on the way we live. According to *Everybody Counts*, the ideas of mathematics affect us on many different levels, among them the following:

Practical—figuring unit prices in the supermarket, understanding the effects of inflation, balancing a bank account, helping the kids with their homework. Imagine how proud Janine felt when she learned to handle her own bankbook! And imagine how Marlene (age about 45) suffered; she was so frightened of numbers that she would not even discuss money matters with her husband, let alone try to balance her checkbook.[11]

False Positives and the AIDS Virus

The test for the AIDS virus is 96 percent accurate. Ten thousand people considered to be at risk were tested for the AIDS virus. John Jones (not his real name) tested positive on the test and two hours later committed suicide. In the note he said that since he had a 96/100 chance of having AIDS, he did not see the sense in going on. His family was shocked.

This shocking, but true, story was cited by Northeastern University professor Margaret Cozzens to illustrate the effects of mathematical illiteracy. She goes on to prove mathematically that most of the people who tested positive in this sample were *not* infected with the AIDS virus. Their test results are called "false positives." Only about one out of ten people who tested positive actually had the AIDS virus. (Margaret Cozzens, Editorial, *Consortium* 32, Winter 1989.)

John Allen Paulos includes a similar analysis of testing for cancer in his book *Innumeracy: Mathematical Illiteracy and Its Consequences* (page 166). He suggests that many such tests are far from reliable, among them the Pap test for cervical cancer and lie-detection tests.

Civic—understanding such public policy issues as tax rates, the education budget, health care, preserving the environment, and the effect of racism, sexism, and other forms of discrimination. How does the Savings and Loan bailout of the 1990s affect my tax bill? What is the meaning of a one-percent rise in the unemployment rate? Which population groups are most affected by this increase? Does the published figure give an accurate picture of the true employment situation?

A graph or a table in a newspaper article often tells us as much in a short space as a whole column of text. Yet many math-fearers pass right over these mathematical representations,

Inflation: Your Shrinking Dollar

Inflation means that your dollar is worth less this year than it was last year. Suppose a half-gallon of milk cost one dollar in 1983. That half-gallon of milk would have cost about $1.35 in March 1991, due to inflation, other factors being equal.

How do I know? Each month government economists figure out the cost of a typical family's expenses, using more or less the same items every month. They choose the cost of living for a certain year, let's say 1983, as the base year. Then they compare the cost for each month with the 1983 cost. Each of these amounts is translated into a number called an index. The base year index, the number for 1983 in this case, is set at 100. The cost in March 1991 turned out to be 135. To use the language of the economists, the Consumer Price Index (CPI) was 100 in 1983 and 135 in March 1991.

I found these numbers in my almanac, listed under "Consumer Price Index." That's how I estimated the rise in the price of a half-gallon of milk.

Can you guess which items cost the most in March 1991, compared with 1983? The three highest items were: education (179), medical care (174), and tobacco products (198). Roughly translated into dollars, college tuition set at $5,000 in 1983 might have risen to $8,950 by March 1991.

The CPI provides the basis for adjusting annual Social Security benefit payments to take account of increases in the cost of living. The increase is called a Cost of Living Adjustment (COLA).

convinced that they cannot understand anything involving numbers. This attitude may be dangerous. These people may be so awed by numbers that they tend to place their faith in anyone who uses mathematical arguments, regardless of the validity of such arguments. Mathematical knowledge helps you to analyze data and make informed decisions about important issues.

Citizen Monitors the Environment

In the summer of 1988 a Washington, D.C., citizen, a lawyer by profession, became concerned about the ozone levels in the area. With the help of his personal computer he processed information that was readily available to the public and then drew the appropriate conclusions about the level of pollution in the environment. No doubt he used his expertise in both law and the environment to do something about the situation. ("All Things Considered," National Public Radio, 26 October 1988.)

Professional—applied to fields ranging from business management and health care to machinist and physicist. Scientists and economists use mathematics to investigate and, hopefully, solve the severe problems that confront the world, among them global warming, AIDS, waste disposal, and war. Both managers and workers must have some understanding of statistics to make sense of the quantities of information they generate and receive. By now there are few fields of work that do not involve computers.

Leisure—games of chance and games of strategy, puzzles, and prediction and analysis of sports events. Recall teenage Alan, in the previous chapter, and his reluctance to play pool because of his fear of math. It may come as a surprise to learn that some people do math because it's fun! Since most people know only the math that they encountered in school, they cannot be faulted for their misconception that school math is real math. Activities such as solving puzzles and playing games are closer to real mathematics than are the exercises to which you may have been exposed in school.

Cultural—"the role of mathematics as a major intellectual tradition, as a subject appreciated as much for its beauty as for its power. . . . Like language, religion, and music, mathematics is a universal part of human culture."[12] All societies throughout the ages have developed mathematical ideas and practices appropriate to their needs and interests.

With mathematics playing such an important role in our society, it is no wonder that many people suffer a loss of self-esteem

Are Women or Men Doing Better?

We know that smoking kills. Yet many of us continue to smoke. The graph shows the percentage of men and of women who smoked cigarettes in three different years: 1965, 1980, and 1988. Obviously both men and women have cut down on smoking. Who has the better record, men or women?

The graph shows that half of men smoked in 1965, while just over three out of ten were doing so in 1988. The proportion of men who smoked in 1988 was just about the same as the proportion of women smokers in 1965. Fewer women smoked to begin with. It seems that the women are doing better.

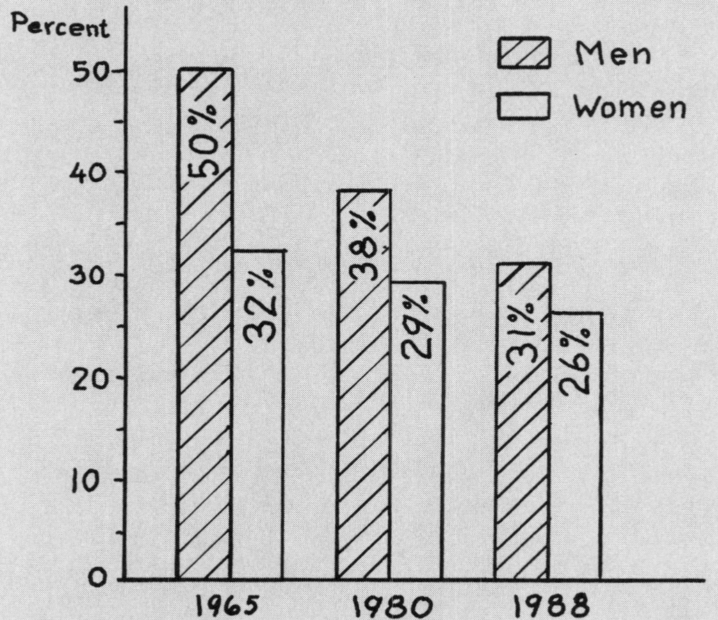

Percent of Population Who Smoked Cigarettes
SOURCE: National Centers for Disease Control, Office of Smoking and Health, reported in *Smoking and Health Review,* March/April 1991.

(*Continued*)

Are Women or Men Doing Better? *(Continued)*

Let's translate the percentages into concrete numbers. For every fifty men who smoked in 1965, thirty-one were still smoking in 1988. In other words, sixty-two out of a hundred former male smokers were still at it in 1988. For every thirty-two female smokers in 1965, twenty-six were smokers in 1988, the equivalent of eighty-one out of a hundred. These figures show that women smokers were slower than male smokers to give up the habit.

In mathematical terms, the ratio of women smokers to the total number of women has always been smaller than that for men. But women have been giving up smoking at a slower rate than men. Both sexes need to improve!

because of a perceived inability to cope with the subject. I had known Becky (age 68) only a few minutes before she confessed to me that she still did not know the answer to "one-half times one-fourth." Then she related the incident that had turned her away from math forever. In seventh grade a male teacher, angry at her momentary lack of attention, had slammed his book on the desk and called her "stupid." Rather than going to the principal as the teacher had ordered, she went home crying. After her parents had complained to the principal, the teacher continued to harass her with remarks to the class such as "she'll cry" and "she'll tell her parents." Becky, formerly an A student in math, just barely passed the subject thereafter. Her father attempted to console her by assuring her that she didn't need more math. "You only need to know up to 12. No recipe calls for more than 12," an allusion to her future role as a housewife. Becky had eight children, taught elementary grades (but not fractions), and then became a librarian. Yet she has never regained her confidence in her ability to do mathematics.

The Current Situation

The words "underserved" and "underrepresented" have entered the vocabulary of concerned educators. "Underserved" refers to the fact that large segments of the population of the United States are not receiving the kind of education that they need and deserve. The "underrepresented" are the groups that are not participating in higher education and in prestigious and well-paid occupations to the degree that their numbers in the population as a whole would warrant.

Who are the underserved and the underrepresented? For the most part, they are the victims of poverty and discrimination on the basis of race, ethnicity, and gender. Let's be more specific.

During the past few years a spate of reports from governmental, educational, and corporate leaders have deplored the low level of mathematics and science achievement in our country. According to the report *Educating Americans for the 21st Century,* the even poorer performance of African-American, Native American, and Latino students "can be traced directly to both blatant and subtle racial discrimination (including stereotyped racial attitudes), and extreme poverty. [When these students] are exposed to a good learning environment, they perform as well as any. Low achievement norms do not reflect ability; they reflect a lack of preparation and early exposure."[13]

These conditions exist in spite of the improvements in educational opportunities of the past few decades. The landmark 1954 Supreme Court decision on school desegregation and the Higher Education Act of 1965 providing financial aid programs for minorities enabled more people of color to participate in higher education. Similarly, the federal Bilingual Education Act of 1968, as well as legal action in the 1970s, contributed to the improvement in the education of Hispanic and other language-minority students. Title IX of the Education Amendments of 1972 prohibited sex discrimination in any educational program receiving federal assistance.

Despite improved education and job opportunities, women, African Americans, and Hispanics are still greatly underrepresented in mathematics-related careers. Black engineers and scientists (including social scientists and psychologists) constituted 2.6 percent of the total number of engineers and scientists in the United States in 1988, up from 1.8 percent in 1978. If African Americans were fairly represented in these fields, according to their share of the population, the number should increase fivefold. Hispanics comprised 1.8 percent, compared with their 8 percent share of the total population. To be fair,

Math Prof's Expertise Helps Farm Workers

Born in Mexico, Luis Ortiz-Franco earned his doctorate at Stanford University and is currently a professor of mathematics at Chapman University in California. After a stint as a research associate at the U.S. Department of Education, he joined the staff of the United Farm Workers of America, where he conducted research for labor law cases, engaged in labor organizing, and participated in labor contract negotiations. His mathematical knowledge was a critical asset in securing fair compensation for the farm workers.

One specific problem involved the union's demands for fair wages and benefits lost during a lockout by a California company. Ortiz-Franco had to estimate the dollar value of the wages lost during the duration of the lockout, the value of the benefits that would have accrued during that period, and interest on these amounts. Through the efforts of the United Farm Workers and its leadership, headed by Cesar Chavez, the workers recovered these funds and negotiated a better contract than the one in force before the conflict. (Personal communication with Dr. Ortiz-Franco.)

more than four times as many Hispanics should be in the pool of scientists and engineers.

It is interesting to note that African-American women were much better represented among female scientists than were African-American men among male scientists—5 percent and 2 percent, respectively. To a lesser extent, Latinas had a larger share than Latinos. Of course, the whole pool of female scientists was very small, only 15 percent of the total in 1986.

Native American representation was about equal to their share of the population. Asians, both men and women, were overrepresented, particularly in the field of engineering, but a large number—27 percent in 1986—were not citizens of the United States.[14]

During the past few years dramatic changes have occurred

in the participation of women in some fields requiring a good math background. In 1979 only one in three accountants and auditors was a woman. By 1986 the percentage had climbed to 45 percent; almost half the people in the field were women. For computer programmers the increase was similar, from 28 percent to 40 percent.[15]

Although few women have entered engineering, the number of female engineering graduates has increased. In 1976 only one out of sixty engineers in the United States was a woman. By 1986 the figure was one out of twenty-five, still woefully small. However, 15 percent of all engineering bachelor's degrees were earned by women in 1985, compared with only 2 percent ten years earlier.[16]

A 1989 report of the American Association for the Advancement of Science, *Science for All Americans,* had this to say about equity in science (including mathematics) education:

> When demographic realities, national needs, and democratic values are taken into account, it becomes clear that the nation can no longer ignore the science education of any students. Race, language, sex, or economic circumstances must no longer be permitted to be factors in determining who does and who does not receive a good education in science, mathematics, and technology. To neglect the science education of any (as has happened too often to girls and minority students) is to deprive them of a basic education, handicap them for life, and deprive the nation of talented workers and informed citizens—a loss the nation can ill afford.[17]

CHAPTER THREE

Myths of Innate Inferiority

> I have learned from experience that my dislike of math was intimately connected to my fear that I could not do mathematics, that I just wasn't smart enough. This is not true for me; nor is it true for anybody else. There are skills which can be learned that in turn offer us opportunities to learn yet more.
>
> Mathematics is not just for genius IQs. We lose a lot as a society when we decide that only the chosen few are capable. Perceived differences in ability are tied in with income, educational opportunities, life experiences, which should give anyone pause when making such judgments.
>
> —*Mary Jo Cittadino, Mathematics Educator, EQUALS, University of California at Berkeley*

Over the years many theories have emerged that purport to prove the innate inferiority of women, Africans and other people of color, and members of the working class. No sooner is the falsity of one theory exposed, than another is put forth to take its place. These sexist, racist, and elitist ideas have had a profound effect on our educational system and on the lives of their victims. Fear of math is one devastating outcome.

Several years ago I received a letter from my friend Jane, asking: "Is it true that girls can't do advanced math? That's what I read in the *New York Times* last Sunday." I was shocked, to put it mildly. Jane and I had been discussing this question for ages, and I was sure that she had no doubts about the ability of women to do any kind of mathematics. What had she read that would induce her to raise this question?

The title of the article, "Girls and Math: Is Biology Really Destiny?," as well as the text, suggested that genetic traits are responsible for the fact that females get lower scores than males on standardized tests like the Scholastic Aptitude Test (SAT) and the American College Testing (ACT) Program, taken by high school juniors and seniors as part of the college admissions process. One researcher, Camille Benbow, was quoted as saying:

"Many people don't want to believe the male advantage could be biological. . . . It took me fifteen years to get used to the idea."[1]

The belief that females have an inborn, unalterable inferiority when it comes to doing math is among the myths that induce them to shun mathematics. Unproven beliefs about the innate intellectual inferiority of any specific group in relation to another affects that group's attitude toward math. Why bother trying to achieve in a field in which one is doomed by nature to be totally incompetent?

In this chapter I shall try to demolish some of the beliefs that still pervade our society, among them:

- Math is for boys.
- Males have more spatial ability than females, and that makes them superior in mathematics.
- Asians are superior to all other groups in their aptitude for mathematics.
- Whites are superior to blacks in intelligence.

Add to this list some myths about standardized tests. Testing is an important issue. It is generally these tests that determine how people are sorted into "achievers" and "nonachievers." These tests often serve to deprive people of access to education and careers. Two myths about testing are:

- Scholastic Aptitude Tests (SATs) measure aptitude or predict achievement in college.
- Intelligence (IQ) tests are a reliable measure of an individual's intellectual capacity.

Later chapters will deal with other myths about the ability to learn mathematics, such as:

- One must have a "mathematical mind" to do math.
- Children who are considered "culturally deprived" or "learning disabled" cannot learn math.

Is Math for Boys Only?

The psychologists Camille Benbow and Julian Stanley are among the main proponents of the point of view that males have a "biological math advantage." In 1980 they made the news with an article in the journal *Science* entitled "Sex Differences in Mathematical Ability: Fact or Artifact?" in which they stated: "We favor the hypothesis that sex differences in achievement in and attitude toward mathematics result from superior male mathematical ability, which may in turn be related to greater male ability in spatial tasks." They arrived at this conclusion by comparing the SAT scores of gifted twelve-year-old girls and boys competing for admission to an accelerated program for "mathematically precocious youth."[2]

From the media's coverage of this issue one got the impression that there was no doubt about women's inferior innate ability in math:

- "Do males have a math gene?"[3]
- ". . . males inherently have more mathematical ability than females."[4]
- "Two psychologists said yesterday that boys are better than girls in mathematics reasoning."[5]

Although many researchers in the field severely criticized the study on which these conclusions were based, the media, in the main, chose to ignore evidence that contradicted these biased views. "It is virtually impossible to undo the harm that the sensationalized coverage has done," declared two leading women mathematicians in an editorial in *Science*. Indeed, within a few short years it became evident that parents' beliefs about their daughters' mathematical abilities were being influenced negatively by the media's coverage of this issue.[6]

Carolyn was growing up during the time of this media blitz. "My mother kept telling me she was never good at math either, and a lot of people told me that girls were better at English and boys were better at math when I was young. So I think this also stuck with me and was in the back of my mind. It contributed to my math anxiety."

More recently Benbow and Stanley attributed the supposed superior spatial ability in males to the effect of the male sex hormone testosterone in the prenatal stage. Supposedly this hormone strengthens the right hemisphere, the part of the brain specialized in spatial visualization. And spatial ability, they claimed, is essential to solve higher-level mathematical problems.[7]

Another Longitudinal Study
by Phyllis R. Steinmann

Once upon a time, in 1940, a research team decided to study the cooking ability of males and females to see if sex had any effect on this ability. The population chosen for the study consisted of 2,000 individuals selected at random. One thousand were males and one thousand were females. Their ages ranged from 5.5 years to 6.5 years. They were observed for an average of two weeks, usually in the kitchen. It was found that both groups were equally adept at spreading peanut butter on slices of bread. Their biggest problem was when the jar of peanut butter was on the top shelf. Then the boys, who were slightly taller than the girls, had a little advantage. The major difference in the sexes was that the girls were twice as likely to put the dirty knife in the sink.

The same subjects were investigated again in 1946 when their ages ranged from 11.5 to 12.5. The large majority of girls, 65 percent, were able to prepare a simple meal, while only 15 percent of the boys had attained this skill.

The final study in the series was done in 1952, when the subjects were around eighteen years old. At this time most of the females, 85 percent, were capable of preparing a full course dinner, but only 5 percent of the males could do this.

The study clearly showed that there was a great difference in the cooking ability of boys and girls. Ten percent of the sample were twins where one was male and the other female. These siblings, brought up in the same environment, showed the same discrepancy in cooking ability. It was decided the only explanation for this situation was that the ability to cook was stronger in female genes than in male genes.

In order to confirm the finding of this study, it was repeated again in 1972 with another group of 1,000 boys and 1,000 girls. At age six, the two groups were again found to be equally able to spread peanut butter sand-

(*Continued*)

Another Longitudinal Study (*Continued*)

wiches. In 1978 at age twelve, they were about equal in their ability to heat up TV dinners in the microwave oven. In 1984, neither group had developed any real cooking skills, opting to eat out at McDonald's.

The results of this study are still being analyzed. One likely conclusion is that in a matter of some thirty years the female genes have been altered, possibly caused by radiation from TV sets.

NOTE: The Benbow-Stanley brouhaha inspired mathematics educator Phyllis Steinmann to write this satire for the Women and Mathematics Education *Newsletter* 6 (Summer 1984). Steinmann also devised Algebra Blox, materials that have proved very useful in her work with math-fearing adults. See "Resources" (Appendix) for a description and ordering information. Steinmann retired from the Maricopa County Community College District, Arizona.

Their hypothesis raises many questions:

- Is spatial ability a prerequisite for doing mathematics?
- Does testosterone have the effect they claim for it?
- Do males really have superior spatial ability?
- Are SAT scores of precocious twelve-year-old children—or even of high school seniors—a fair criterion for judging mathematical ability?

Spatial Ability

Let's first examine the notion of spatial ability. There is no general agreement among psychologists, mathematicians, and mathematics educators as to what constitutes spatial ability. To some it means the ability to visualize a three-dimensional figure in various positions. Others broaden the definition to include the capacity to handle any type of task involving relationships in space. Almost everyone has this ability to at least some degree.

We can recognize that one object is larger than another. We might be courting disaster if we failed to judge the closeness of moving vehicles as we cross the street. Anyone who plays ball or threads a needle has some intuitive feeling about spatial relationships. Experienced pianists know exactly where to place their fingers without having to look at the keyboard, a facility that comes with practice.

Is spatial ability a prerequisite for doing math? Not necessarily. Some branches of mathematics do not involve spatial relationships at all. In some other branches, problems can be solved by either spatial or purely verbal methods. Two investigations involving the solution of a variety of problems were carried out with junior and senior high school students. The female and the male students had each been divided into two groups, high-spatial and low-spatial ability, on the basis of their scores on relevant tests. The low-spatial males did somewhat better in solving the problems than did their high-spatial counterparts. In some aspects of the solution process the high-spatial females outperformed both male groups. Only one group performed significantly more poorly than the other three. Females with low scores in spatial skills seemed to have much more difficulty in doing mathematics in general. This may be important because even a small group of females with low achievement levels can pull down the average score for females overall.[8]

Much has been made of right-brain and left-brain differences.[9] In females the left hemisphere is supposedly more developed, resulting in greater verbal ability, while men, with their right brain superiority, are thought to excel in spatial ability. Tests show, however, that both halves of the brain are involved in most mental processes. In one study participants were given two tasks, one requiring spatial judgment and the other, numerical judgment. Using an array of recording electrodes on the subject's scalp and a pattern-recognition computer program, the researcher determined that neither task involved just the left side or the right side of the brain. On the contrary, each task produced a complex pattern of electrical activity on both sides of the brain.[10]

As for the effect of the hormone testosterone, no one has been able to show concrete evidence for its influence on brain functioning in the prenatal stage. In fact, any hypothesis involving genetic influence on intellectual ability is a "black box" situation. It's just not possible to prove it one way or the other. We do know, however, that sex differences in mathematics achievement, outside of the SATs and similar tests, have by now become almost negligible. Obviously, genes don't change from one generation to the next. The improvement must be due to social factors.[11]

But it's not necessary to rely on untestable theories about genes and hormones to compare male and female spatial ability. A variety of tests are available to measure different aspects of this skill. On timed tests, males have a slight edge, chiefly on exercises involving mental rotations of a three-dimensional object. But when given adequate time, women do just as well as men. Critics raise objections to the use of such tests to conclude that women are innately inferior in spatial ability. For one thing, most of the tests are timed, and many women don't do as well as men on timed tests. Secondly, the exercises are not placed in a meaningful context, a factor that seems to be more vital to women than to men.[12] I will discuss these factors later in this chapter and in subsequent chapters.

At present we have relatively little information about the interrelationships of brain, ability, and learning experiences. But we can point to studies that show the positive effects of training. To cite one example, Joan Ferrini-Mundy gave women in her calculus classes specific training in imaging the sketching of a solid figure as it revolved about an axis. She concluded that "practice on spatial tasks enhanced women's ability and tendency to visualize while doing solid-of-revolution problems." A further outcome was an improvement in women's performance to the point where they equaled or surpassed the men in the calculus courses.[13]

A study conducted by the Educational Testing Service (ETS) confirms the importance of training and experience in developing the spatial skills of both females and males. ETS compared the spatial abilities of high school seniors in 1980 with those of 1960 seniors, using a test called "Visualization in Three Dimensions." Not only did the 1980 group score lower than their 1960 counterparts, but their scores were equivalent to those of first-year high school students in 1960. Although in 1980 males continued to perform better than females, the gap between the sexes had narrowed. These outcomes are a strong indication that experiential factors influence the development of spatial ability. ETS researcher Thomas Hilton believes that the decline is due to the way young people spend their time. When they are

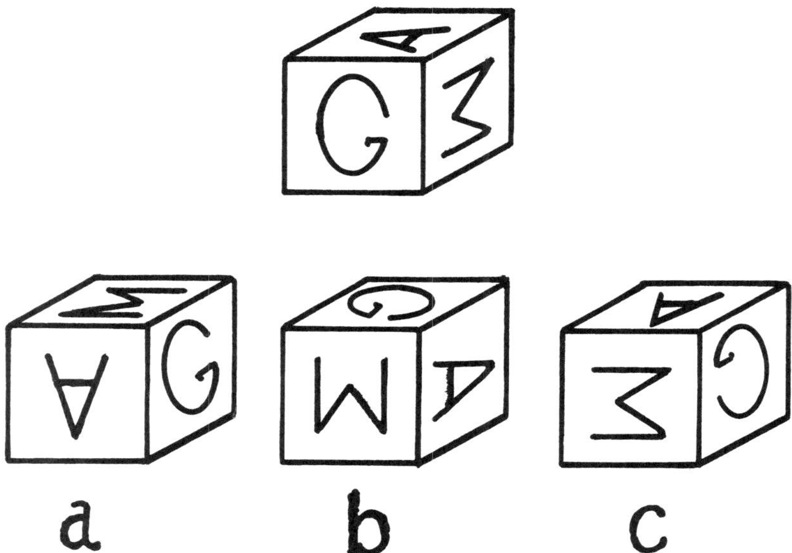

Which Block in the Bottom Row Matches the Block on Top? (b)

engaged in watching television, they have less time to spend in sports and hobbies like constructing models. Schools, moreover, are offering fewer courses in three-dimensional geometry, mechanical drawing, sewing, art, wood shop, and metal shop. As another researcher remarked, although these tests don't measure everyday uses of three-dimensional abilities, the decline in skills is alarming.[14]

From time immemorial women have been engaging in tasks that require spatial ability. Our foremothers sewed and knitted clothing and household items, as many women continue to do today. The patterns in Dineh (Navajo) rugs, Hopi baskets, and Nigerian painted resist cloth, to name but a few examples of women's art, call for the application of high-level spatial skills. Ndebele (South African) women paint elaborate geometric patterns, usually of their own design, on the walls of their homes.[15] All these women work with no reference to sketches or measurements. Instead, they rely on their ability to visualize.

Many observers have remarked about the extraordinary talent of Inuit (Eskimo) women and men for spatial orientation and visualization, a skill required by their vast, snowy, virtually featureless environment. Studies find that women athletes consis-

Dineh (Navajo) rug. Courtesy of Vestal Public Library. (Photograph by Sam Zaslavsky.)

Dineh (Navajo) rug. Courtesy of Sydra Schnipper. (Photograph by Sam Zaslavsky.)

tently do better than average men on spatial-skills tests. Were all these people born with this facility, or is it more logical to assume that the skills were developed in the course of years of practical applications?

In the sphere of work in the industrialized world, few occupations require the type of spatial ability in which women don't test as well as men. Yet these tests are used as screening devices to exclude women from careers and educational programs, even from those that don't require such skills.

Scholastic Aptitude Tests and Mathematical Ability

The Scholastic Aptitude Test (SAT), as its name implies, is supposed to measure a person's aptitude. The SATs are multiple-choice tests, one designed to measure verbal aptitude and the other, mathematical aptitude. Scores on each test range from a low of 200 to a high of 800. In recent years the SATs have come under considerable criticism. In a scathing attack on the parent organization, Educational Testing Service (ETS), David Owen points to major weaknesses in the results obtained from these tests.

For one thing, the multiple-choice—critics call it the "multiple-guess"—format requires the test taker to choose one correct answer, working within a tight time limit. Women tend to take more time than men to consider the alternatives and reflect upon their choices. As a result, they may complete fewer items. Socialized to follow rules, they are apt to take more seriously the warning about penalties for wrong answers and to omit questions when they are not sure about the answers. Often it's the bold risk taker, the aggressive guesser, who gets the high score. Women don't do as well as men on this type of test, nor is this the kind of aptitude that college work demands.[16] Of course, these comparisons between men and women only indicate a tendency. They certainly do not apply to all women and all men. Some women are not afraid to take risks, while some men work slowly.

Pam (age 20) entered a community college after several years of bad experiences with high school math. "I felt incredibly lacking in intelligence." In college she learned from a sympathetic instructor that math problems do not have to be done so fast, and that a wrong answer is not a calamity. "He explained that time is not defined when it comes to math problems. What

Timed Tests

In a poignant passage of her book, *Balm in Gilead,* Sara Laurence Lightfoot tells about her mother's frustration with timed tests. At the age of fourteen Margaret Laurence moved to New York City to attend single-sex Wadleigh High School. Sara Lightfoot writes:

> In order to be admitted to Wadleigh, prospective students had to take entrance examinations—both achievement tests and departmental exams. For the fourteen-year-old girl from Mississippi, the achievement tests were an ordeal—a strange format for testing knowledge at a baffling, accelerated pace. She did poorly on achievement tests (a pattern that would continue through college). "I was not geared to rapid test taking. I expected to sit there and think about it. I remember them taking the test away before I ever got started." The contrast in style and pace between the teaching methods in her black school in the Deep South and in the classical, academic high school in the urban Northeast must have contributed to Margaret's frustrations and low scores on the standardized tests. But her depressed scores did not simply reflect the differences in cultural style and educational expectations. Margaret's intelligence does not lie in the quick, discrete response, in the immediate recall of fact. Her mind works more slowly as it considers various perspectives, considers competing interpretations, and puts pieces together. . . . [Her] intellectual and temperamental tendencies did not make her a good candidate for the standardized tests.

Margaret Laurence's intellectual and temperamental tendencies did help her to earn both a medical degree and a public health degree, and to pursue a distinguished career as a psychiatrist. (Sara Laurence Lightfoot, *Balm in Gilead* [Reading, Mass.: Addison-Wesley, 1988], 84.)

an idea! Finally I was able to see that taking longer on a problem doesn't necessarily mean that I will never find the answer or that I'm not *capable* of finding the answer. I am." Hardly the mind-set for taking a timed SAT test, but it serves her well in her second course in calculus.

If the SATs were truly measures of innate aptitude, would coaching improve the scores? Coaching is now a multi-million-dollar business, and many high schools, particularly in high-income areas, have established their own coaching classes. Coaching schools claim to raise the score on each test by up to 100 points. One of my own students, on his second try at the math SAT, was able to better his score by almost 200 points, with my help and several months of practice. Although the ETS guide discourages guessing, and wrong answers are penalized, coaching schools teach "smart guessing" as a test-taking technique. It often works wonders.[17]

Taking advanced math and science courses in high school also serves to increase scores on the math SAT. In 1992 there was a spread of 146 points between those who had taken over four years of high school math and those who had studied the subject for only two to two-and-a-half years. Fewer women than men take advanced math and physics, although the gap is becoming smaller. And as Patricia Campbell points out: "Low-income students and students of color who take algebra and geometry go to college in numbers equal to wealthier whites. However, only half as many low-income students and students of color take these important courses."[18]

Because the SATs have been criticized on the grounds of culture and gender bias, ETS has instituted an internal review to eliminate bias in the questions. The major source of bias, however, cannot be removed by ETS. It lies in the inequities in our society. Wealthy students and children of highly educated parents score far higher than children of low-income and poorly educated parents. In 1992 there was a difference of more than 100 points in the average math scores of low-income and high-income test takers. In the lower-income categories many more women than men take the test; men may opt for the military or a job instead of college. The preponderance of low-income women with inadequate math backgrounds tends to bring down the average score for women. A similar spread is found on the basis of parents' educational level.

Asian Americans have the highest math SAT scores, higher

SAT-Mathematics Averages by Family Income, 1992

Family Income	Percent SAT takers	SAT-Math average	Percent Female	Percent Male
Less than $10,000	5	415	61	39
$10,000–$20,000	11	434	58	42
$20,000–$30,000	14	453	55	45
$30,000–$40,000	17	467	53	47
$40,000–$50,000	13	480	51	49
$50,000–$60,000	11	492	51	49
$60,000–$70,000	8	501	50	50
$70,000 or more	20	530	49	51

SOURCE: *College-Bound Seniors: 1992 Profile of SAT and Achievement Test Takers* (New York: College Entrance Examination Board, 1992).

than those of whites, although their verbal scores are lower than whites'. Their achievements in math and science have led to stereotypes about innate mathematical ability, stereotypes that Asian students frequently resent and try to counteract. Many of these students come from highly educated professional families, recent immigrants to the United States. They may have attended schools in their native countries that are tougher than ours, and they are accustomed to spending time on homework rather than on TV. Parents may exert a great deal of pressure for their children to succeed.[19]

Here are the comments of two Asian high school students:

> *Chu (from Hong Kong)*: "I would say that we study math at a higher level in my country. The entire system is different." He adds that Asian students are attached to math and science because it is an international language. "It's just a way for survival . . . a way to move ahead. Parents think that kids should go for science in order for them to secure a better job."

> *Grace remarks about her parents:* "They feel that math and science is the key to success in the world." She admits that being stereotyped irritates her and unconsciously makes her want to do poorly in science and math to spite everyone.[20]

Working-class students of both sexes score lower than middle-class and wealthy white males. A disproportionate number of these young people are members of minority groups. In 1982 the College Board publicized for the first time a breakdown of SAT scores by race, revealing that African-American high school seniors scored about 100 points lower than the national norm in mathematics. The vast difference in average family income provided an obvious explanation. The scores of all students rose with family income and parents' educational level, and the gap between the scores of black and white students narrowed as income rose. The telling point is that the average family income of white families was about twice that of black families.[21]

The noted African-American educator, Dr. Kenneth B. Clark, commenting on the College Board's action, pointed to a more insidious factor:

> The traditional "explanation" for blacks' consistently low academic achievement scores is that blacks are inherently inferior intellectually. This "explanation," in its subtle and gross forms, is a manifestation of the racism that lies at the core of, and perpetuates this educational and social problem.
>
> Black children are educationally retarded because the public schools they are required to attend are polluted by racism. Their low scores reflect the racial segregation and inferiority of these schools. These children are perceived and treated as if they were uneducable. From the earliest grades, they are programmed for failure. Throughout their lives, they are classic examples of the validity of the concept of victimization by self-fulfilling prophesy.[22]

ETS no longer claims that the tests actually measure aptitude, but states that they are useful in predicting college performance. The argument doesn't hold water. Studies have shown that high school grades are better predictors than the SATs. For example, women's grades in high school math classes and subject matter tests, like the New York State Regents exams, have been comparable or superior to those of young men. And women do as well as men in college math and science courses, even though their SAT scores are lower. An Educational Testing Service study of almost 47,000 college students, matched for mathematics courses and grades earned, revealed that males had averaged thirty-three points higher than females on the math SAT. But

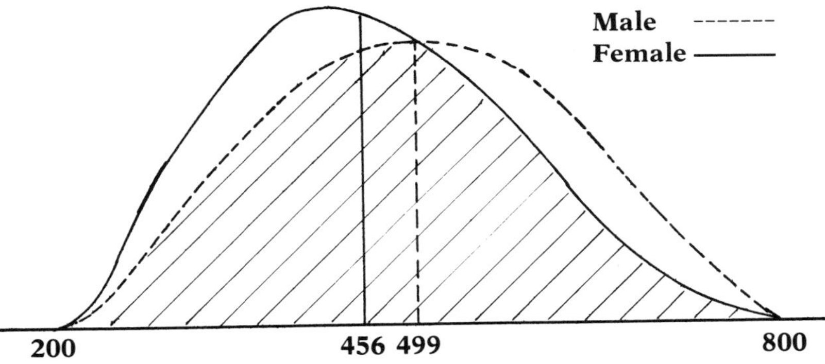

Distribution of SAT-Mathematics Scores by Gender
SOURCE: *College-Bound Seniors: 1992 Profile of SAT and Achievement Test Takers* (New York: College Entrance Examination Board, 1992).

because these SAT scores are used as a basis for scholarship awards, as well as for college entrance, women lose out.[23]

In sum, SAT scores reflect education, attitudes, experience, and test-taking skills. Yet we are led to believe, years after the exposure by Clark, Owen, and many others, that the tests actually measure innate ability. In an Op-Ed in the *New York Times* in 1989, Steven Goldberg, chair of the sociology department of the City College of New York, argued: "It is the male superiority at various spatial and mathematical aptitudes that correlate with college performance. . . . No serious researcher questions male superiority in mathematical reasoning." His argument rests on the difference in mathematics SAT scores of men and women. He maintains that the "SATs are . . . the best single predictor of college performance" and that they "emphasize intelligence." Subsequently the *New York Times* published several letters, including one from the mathematics chairmen of Harvard and Princeton universities, refuting Goldberg's arguments.[24]

IQ Tests and Intelligence

The IQ test is popularly thought to be an objective measure, like height or weight, that can be used to arrange children according to their degree of intelligence.[25] Nothing is further from reality. Long before the existence of the SATs, the IQ test served as a sorting device.

Most of us women scored as well as most of you men, and we get better grades in school...
Yeah, but we get more scholarships!

Tests to measure "intelligence" were devised in France in the early 1900s by Alfred Binet, to identify children who might profit from special education. He assigned an age level to each task, based on its degree of difficulty. To compute the intelligence quotient (IQ), the mental age was divided by the chronological age of the child.

American psychologists seized upon Binet's concept. They took Binet's scores as measures of a fixed entity called intelligence which, they assumed, was largely inherited and marked people for their station in society. Stanford University professor Lewis M. Terman adapted Binet's tasks for mass testing. These tests have had a sinister effect upon the history of our country. They have been used to track children in school, to limit the access of African Americans to higher education, and to justify racial segregation. They contributed to the passage of the Immigration Act of 1924, which imposed strict ethnic quotas that excluded most eastern and southern Europeans, and which was not repealed until 1965.

Arthur Jensen, in his 1969 article "How Much Can We Boost IQ and Scholastic Achievement?" sought to prove that blacks' lower scores on tests were due mainly to heredity, and could be modified only very slightly by experience. Blacks, he argued, are capable of little more than rote learning.[26]

Most of the media picked up Jensen's thesis uncritically, just as they have devoted major attention to Benbow and Stanley's conclusions. Imagine the effect on a young African American reading an article with the headline "Born Dumb?" Although children's advocates have forced the banning of mass IQ testing in many districts, the damage has been done. No doubt many educators use Jensen's arguments to confirm their stereotypes about the ability of their black students.[27]

IQ-type tests are widely used today, although they are more likely to be called "aptitude" tests. They serve to track a disproportionate number of students of color and working-class children into low-level or special-education classes, thereby depriving them of opportunities for productive and creative lives.

Some testing proponents favor the use of IQ-type paper-and-pencil tests for screening applicants for all kinds of jobs, claiming that the score on such a "cognitive ability" test is a valid criterion for selecting employees for *all* jobs, including such skill-oriented work as welding. The Lawyers' Committee Employment Discrimination Project has challenged the use of such tests, alleging that they are used to get around affirmative action guidelines for hiring members of minority groups. Many candidates have adequate skills for jobs like welder or beautician, but have been deprived of the educational opportunities that would enable them to do well on written "cognitive ability" tests, thus excluding them from these positions.[28]

The practice of IQ testing rests on several questionable assumptions: that every person has a specific, unalterable entity called intelligence, that it can be measured by a test, and on that basis it can be assigned a valid number. Many of the criticisms of the SATs as a measure of innate aptitude apply as well to IQ tests. They are clearly biased toward the groups that are most successful in our society.[29]

On the subject of bias within the tests themselves, Jonathan Beckwith of Harvard Medical School writes:

> The elimination of bias is an enormous problem in the design of tests. Consider the history of the Stanford-Binet IQ test. When it was first prepared in 1916, women

> had higher scores than men. Since the test makers apparently made a judgment that women and men were equal in intelligence, the tests had to be redesigned. Certain questions in which women did consistently better than men were eliminated, so as to result in the same mean score for men and women. The same people could also have decided, if they had wished, to readjust tests in order to have blacks and whites have equal scores. What is clear is that assumptions about the relative intelligence of groups of people have gone into the very design of the SAT and other tests.[30]

Many studies have demonstrated beyond doubt that intellectual capacity improves significantly with training, education, and richer opportunities.[31] To cite one example, each year over a period of nine years, a number of African-American students from the public schools of Washington, D.C., were randomly chosen to receive scholarships to an innovative independent high school. Although these students worked, many were failing math and science. This school stressed the use of language and higher-order thinking skills, in contrast to the rote memorization methods the students had been taught previously. With special courses to enhance their thinking ability and with additional attention from encouraging teachers, these young people were able to make tremendous strides in their math and science work.[32]

Israeli psychologist Dr. Reuven Feuerstein, an adamant opponent of traditional IQ tests, has developed an alternative testing instrument based on an entirely different model of intelligence. He believes that learning *potential* is the most important feature of intelligence, and emphasizes the key role of experienced adults, who, through active intervention, can help children learn to think, to solve problems, and to overcome obstacles. In his Learning Potential Assessment Device, a substitute for the IQ test, a trained examiner gives hints and encourages reflection to enable the subject to arrive at solutions.

Feuerstein and Jensen are at opposite poles in their views of intelligence. On the basis of years of work with thousands of children in several countries, including displaced Jewish children traumatized by their experiences during and after World War II, Feuerstein asserts that thinking skills can be taught. His

Instrumental Enrichment program of "Let Me Think" exercises is designed to do just that. The teacher's role is fundamental. Teachers must sincerely believe that *all* children can achieve, while at the same time playing a crucial role in helping students develop their cognitive skills.[33]

Needed: Different Research and More Action

The stereotypes about the capabilities of women and of people of color have been with us for centuries. They have been used to keep women "in their place" and to justify slavery, conquest, and exploitation. But it is only in recent times, with the development of statistical methods, that the stereotypes have been quantified and thus made "scientific." People may be impressed when they read that many thousands of students were tested in the Benbow-Stanley study. How could the conclusions possibly be wrong when so many subjects were involved?

Furthermore, readers of these sensational articles are led to believe that all males have better math scores than all females, or that all European Americans have higher IQs than all African Americans. Nothing can be farther from the truth. It would be more accurate to point to the large overlap in the scores of the two groups under comparison—assuming that one puts any faith in scores on IQ tests, SATs, and other standardized tests as a measure of innate ability and intelligence. And many of these differences in average scores have been growing smaller in the past few years.

Reports such as those by Benbow and Stanley and by Jensen have a profound effect upon educational policy at all levels. It is of crucial importance that all people develop the tools that can enable them to analyze such reports critically—to look at the assumptions made by the investigators, to examine their use of numbers, to judge whether the conclusions are warranted.

Why is it that the proponents of racist and sexist points of view are able to get the most attention? Why don't we see headlines that proclaim:

- "Everyone Can Learn to Think!"
- "High Income Blacks and Whites Have Similar Scores"
- "Girls and Boys Have Equal Aptitude for Math"

When "More" Doesn't Mean "Better"

Yves Christen, described as a "prolific science writer with a doctor's degree," discussed the work of Benbow and Stanley in his book, *Sex Differences: Modern Biology and the Unisex Fallacy.* Obviously he was so impressed by the number of subjects in their studies that he ignored completely the validity of the testing methods and the objections raised by researchers in the field. Here is a passage from his book:

> Their research was conducted first on 10,000 children in the Johns Hopkins area, and subsequently on a sample of 40,000 Americans! As of the publication of the first part of their work, the results were clear: more boys than girls get good grades in math. Better still, or rather worse, as one ascends the hierarchy of mathematical ability, the differences become more and more overwhelming. Faced with increasingly difficult problems, even relatively high-scoring girls fade away. Intelligence tests that examine mathematical aptitude confirm this major gap. . . .
>
> Three years later . . . their sample included 40,000 people, and the results were even clearer with regard to differences at the highest levels of ability. Consider SAT (Scholastic Aptitude Test) scores of 420 and higher: the ratio of male to female success is 1.5. . . . But at the 700 level, the ratio reaches 13. . . . After this, is it still possible to speak of chance and cultural influence? Most experts find that difficult to believe.

(Yves Christen, *Sex Differences: Modern Biology and the Unisex Fallacy* [New Brunswick, N.J.: Transaction, 1991], 72. Translated from the 1987 French edition.)

Studies have demonstrated that all of these statements are true, but the media don't consider these stories newsworthy—read "sensational." A more significant reason is that such views threaten the power structure as it exists today—predominantly white, male, and wealthy. When women and people of color demand equal rights and affirmative action, they can be confronted with "scientific studies" proving that biology limits their capabilities. "No use trying, folks, because you just haven't got what it takes!"

For example, proponents of an act to provide equal opportunities for women in science and technology were told by opponents of this legislation that it was a waste of money because women are genetically inferior when it comes to mathematics.[34] When Jensen argued that blacks were genetically less capable of learning than whites, and that better educational programs would hardly improve their performance, some government officials seized eagerly upon this argument to urge the abolition of compensatory programs.[35]

In a review of studies about women's mathematical aptitude, researcher Susan Chipman concluded: "Sometimes I fear that what makes research on sex differences so sexy a topic, such front-page material, is the lurking tendency to see in it justifications for social policies of exclusion and differential treatment that most of us hope are *passé*. Small average sex differences are important because. . . . Why?"[36]

With the growth of technology, the differences between the educated elite and the neglected victims are becoming greater. As mathematics educators Arthur Powell and Stuart Varden comment: "Unfortunately the commonly felt view (though never stated in public) that computers and related aspects of the new technology are too complex for working class blacks, Latinos, and women has not only locked them out of the computer field, but also has caused them to internalize this myth."[37]

Interestingly, a test conducted at Claremont Graduate School to measure computer aptitude, interest, and knowledge revealed that women's aptitude was equal to that of men. This was an unexpected outcome for the test developers, who had already written in their manual that women had less aptitude! At all levels of computer courses, women's performance was on the same level as that of men.[38]

Meanwhile the myths of innate inferiority persist, in spite of research studies disproving them. Many school systems continue to track students on the basis of IQ and "aptitude" test scores. Funds for improving educational opportunities and methods are grossly inadequate; why waste money on students who are incapable of learning?[39]

Far more in-depth research is needed on the environmental factors that affect learning. Mathematics educators Laurie Hart Reyes and George M. A. Stanic call for a research program to analyze the ways in which race, gender, and socioeconomic status influence the learning of mathematics: "Clearly, we live in a society where racist, sexist, and classist orientations exist in institutions and individuals. What is not clear is how such ideas are transmitted to and through schools, how ideas are mediated by democratic ideals of equality and equality of opportunity, and the extent to which teachers and students accept and resist the ideas. More specifically, we do not yet fully understand how these ideas affect the teaching and learning of mathematics."[40]

The victims of racist, sexist, and classist policies know from their own experiences that these policies have held them back. Research alone will not solve their problems. Research results, when put into practice, can, however, shift the blame from the victims and win support for campaigns to establish more equitable policies and funding priorities, leading to equal educational outcomes. The real issue is justice for all groups in all aspects of life. Only when such justice is attained can we begin to investigate the possibility of inherent racial, ethnic, or gender differences. Differences, not inferiority.

For those who have been victimized by stereotypes about innate inferiority, it is reassuring to realize that it is not too late to make up for lost opportunities. Everyone can learn math at any age, and the more one works at it, the more skilled one becomes. We know now that positive experiences stimulate brain growth throughout our lives. Instead of "biology is destiny," we can say with assurance that our choices may determine our biology![41]

As Spelman College president Johnnetta Cole wrote: "The reason there are only a handful of African American women physicists is not because African Americans are dumb and women can't do math!" In the following chapters we explore these reasons.[42]

CHAPTER FOUR

"A Mind Is a Terrible Thing to Waste!" Gender, Race, Ethnicity, and Class

> I attended high school . . . and set as my goal that of my elder sister, Rosemarie. Upon graduation, I would be employable as a secretary. I considered this ambitious, setting high sights for my future.
>
> It meant, however, that in place of mathematics or science courses throughout my four years of high school, I took shorthand and typing courses. . . . Going to college was like suggesting I could go to the moon. An English teacher in my sophomore year suggested just that to me one day following class! She asked if I'd ever considered college, encouraging me by saying that she thought I'd make an excellent teacher. The seed was planted, but I proceeded with my original plan formed before high school—to graduate and get a job as a secretary. . . . Decisions about what courses to take in high school were pragmatic and realistic given [my family's] financial situation.
>
> —*Mary Jo Cittadino, Mathematics Educator,*
> *EQUALS, University of California at Berkeley*

The call is out for well-educated workers, people with technical backgrounds, with college degrees, and with graduate-level education. Out of necessity, government, industry, and institutions are looking to women and people of color to fill these job slots. These people will need to know mathematics. Why are so few of them prepared to fill the positions of the future?

Studies show that Americans really do believe the myths about innate inferiority. Not only do they think that some individuals "don't have a mind for math," but that whole groups of people—women, poor people, people of color (except Asians)—

can't learn mathematics. The sad part is that the victims themselves have internalized these beliefs and tend to give up before they have given themselves a chance.

Cultural factors are significant in young people's choices. Girls and boys are assigned different roles from earliest childhood, and peer pressure reinforces the belief that "math is not for girls." Young African Americans may think of mathematics as something that only white people do. The traditions of some groups in our society clash with the demands of formal schooling. As a result, capable people are disbarred from participation in many careers.

More potent than gender, race, or ethnicity in determining one's future is the factor of poverty. Poverty affects people of color disproportionately, and is often a direct result of discrimination in employment on the basis of skin color, language, or culture.

Let's examine these factors more closely to see how they have affected some people's lives. In this chapter we'll consider the influence of the home and community. The next chapter will deal with the educational establishment.

Socioeconomic Status and Mathematical Potential

In 1990 one in five children in the United States under the age of eighteen lived in poverty. The majority of these poor children are white. For children of color, the figures are grim—almost half of black children and two out of five Latino children live below the government's "poverty line." Since 1979 white and Latino poverty rates rose faster than the rate for blacks. Poverty rates for American Indians are equally grim.

Most poor people live in the inner cities, in small towns, and in the rural areas of our country. Schools may be inferior and social services limited. Many adults in these communities are high school dropouts, eager to help their children to succeed, but unaware of just how to go about it. They cannot afford the perks that middle-class parents give their children—good preschool care, out-of-school enrichment programs, summer camps, tutoring services, to name just a few. Gone are the potential role models, the successful professionals and business people; they have moved to more prosperous communities. Most devastating of all,

Who Are the Poor?

Every year the United States Census Bureau issues a report about the people of various racial/ethnic groups living in poverty. Many economists believe that the government sets the poverty line too low. However, I will use the Census Bureau's figures.

The table gives the numbers and percentages for the year 1990. I have rounded all the figures to the nearest whole number. You might wonder why figures for Hispanics are listed separately. The Census Bureau classifies Hispanics as belonging to any race; Hispanics actually are counted twice in the table. The category "Other" includes Asians, Pacific Islanders, American Indians, Eskimos, and Aleuts, to use the Census Bureau's classifications. The population figures are probably higher than those listed here, particularly for blacks and Hispanics, due to the undercount during census-taking—but that's another story.

Poverty in the United States in 1990

	White	Black	Other	Total	Hispanic
Population (millions)	209	31	9	249	22
Number in poverty (millions)	23	10	1	34	6
Percentage of group	11%	32%	12%	14%	28%

Now let's do some comparisons. Look first at the percentages, the last row in the table. Blacks had the highest percentage of people in poverty, 32 percent, about three times the 11 percent rate for whites. One out of every three black people was poor in 1990. The Hispanic poverty rate, 28 percent, was close to that of blacks. We'd have to dig deeper to find out the poverty rates of the various groups included in the category "Other."

(*Continued*)

Who Are the Poor? (*Continued*)

Which group had the largest *number* of people in poverty? The middle row tells the story. In spite of the widely held impression that most poor people are people of color, we see that two out of every three poor people were white in 1990. But it would be incorrect to conclude that poverty is a greater problem for whites than for people of color. The problem is three times as severe for blacks as for whites.

Three important statements about poverty in the United States in 1990 (the prevalence of the number "three" is a coincidence):

- One out of every three black people was poor.
- Two out of every three poor people were white.
- The rate of poverty for blacks was three times the rate for whites.

the environment and experiences of poor people tell them that they are not worthy, that society doesn't need them or care about them. It is to their credit that many of these children succeed against such great odds.[1]

In 1983 the National Science Board published a well-researched report, *Educating Americans for the 21st Century,* proposing a plan of action with the grandiose goal of "improving mathematics, science and technology education for all American elementary and secondary students so that their achievement is the best in the world by 1995." We know now that the goal is utopian, but the contents of the report are quite down to earth. In the section on quality education for all students, the document discusses students of different races, students whose parents do not speak English in the home, and students from disadvantaged socioeconomic backgrounds: "Indeed, the low achievement scores of a significant proportion of such students can be traced directly to both blatant and subtle racial discrimination (including stereotyped racial attitudes), extreme poverty, and, in some cases, unsatisfactory rural or urban conditions."[2]

All children, regardless of social class and racial or ethnic origin, enter school with some mathematical knowledge, a desire to learn, and the ability to perform adequately. A study of children in Baltimore concluded that black children and white children entering first grade have almost identical skills in computation and mathematical reasoning. The investigators examined the effects of race, parent education, social class, kindergarten attendance, parent expectations, and family type (one- or two-parent home). Parent expectation was the single most important factor influencing children's achievement. Children whose parents expected that they would do well actually scored significantly higher in math reasoning. Associated with this factor was parent education. Children whose parents had graduated from secondary school did better than children of parents who had dropped out.[3] Since high school dropouts are more likely to be poor, the connections are obvious.

Marsha (age 44) is finally completing her last year of college. She had avoided math for years after having had difficulty with Algebra I. "My mother and father did not to go to high school, so education was valued—to a point—but honest work was valued more. My parents did not have advanced math so could not understand its value to encourage me to learn more than what was necessary to get by."

The experiences of a small group of African-American seniors in a Chicago high school bring home vividly the effects of poverty. Chosen on the basis of teacher recommendations to participate in a class for the gifted—their test scores were too low for them to qualify on that basis—they must overcome innumerable hurdles: "Where will I get the fifty cents for bus fare to get to school tomorrow? Can we separate Elliot, the trigonometry ace, from his street gang? Will Theo, the calculus whiz, get into a college that will also accept his girlfriend and their baby?" These young people must face the hostility and resentment of their peers for being different. Most of their parents are high school dropouts, eager to have their children continue their education, but with little know-how about the requirements.[4]

Young African Americans in Memphis public housing expressed their resentment about the difficulties they face. A fifteen-year-old student spoke about her high school: "Other schools call it a 'project school.' They say stuff like we can't go anywhere without a police escort. They call us 'at risk.' When

The Rich Get Richer, the Poor Get Poorer

How would you like to receive an annual income of $99.6 million? True, that sum was shared by two men in 1990, the chief executives of Time Warner, and it includes stock options as well as cash. Life had not always been so good to the bosses. In 1975 they were making about thirty-five times the salary of the average worker. By 1989 the ratio had climbed to 120 to 1. While executive pay was soaring in the 1980s, the wages of manufacturing workers, after adjustment for inflation, declined by about 10 percent.

You might want to compare your annual income to the $50,000,000 bonanza received by each Time Warner chief executive officer. Be sure to include all the zeros, the most important part of the number.

To sharpen the impact of the statistics about wealth in our country, imagine that the entire population of the United States can be represented by one hundred people. Of these one hundred folks, at least forty at the lower income end suffered declining earnings during the period 1977–1989. In 1989 the richest person—just one person of the hundred—had an income of close to one-half *trillion* dollars, equal to all the federal tax revenues. This fact inspired Cornell University economist Robert Avery to joke: "If they [the richest one percent] are taxed at 100 percent, everybody else can be taxed at zero." (Steve Lohr, "Recession Puts a Harsh Spotlight on Hefty Pay of Top Executives," *New York Times*, 20 January 1992, A1; Sylvia Nasar, "The 1980's: A Very Good Time for the Very Rich," *New York Times*, 5 March 1992, A1.)

they stick these tags on us, we end up doing exactly what they want us to do: fail. But I'm not at risk. I'm not going to fail."

The principal of this school works overtime to help her students succeed. "We're in an area that is terribly economically depressed. Our kids aspire to the same things that other students aspire to. But some of them don't know how to get there."[5]

Women and Math: The Influence of Family and the Community

Why are many American women lacking in confidence in their ability to do math? Not only American women, but British, Australian, and others—but not all women. In a society that believes in women's innate inferiority when it comes to doing math, the very air they breathe tells girls to give up. Many factors have contributed to this lack of confidence. Although girls have been doing better in recent years—in fact, on most measures outside of the SATs, girls perform as well as boys—women who are now adults have been strongly influenced away from mathematics.

Much of the research on women cited in the next few pages pertains to white females.[6] Latina, African-American, Asian-American, and American Indian women may have had quite different experiences, as their stories will reveal.

Girls start out ahead of boys in talking, reading, and counting. In elementary school they perform better in computation, and they earn equal or better grades at all levels. Yet boys and girls have different experiences at home, in school, and in the wider world, a difference that becomes crucial in their future choices.

Many parents expect boys to be aggressive, self-confident, and independent, while girls are socialized to be docile, well-behaved, and dependent. Young boys have all sorts of toys that encourage exploration and problem solving—sets of blocks, chemistry and construction sets, games and puzzles. Video games, television, and films portray aggressive male figures and passive female figures that serve as role models for youngsters. Boys are more likely to engage in athletics, an excellent way to develop spatial ability. They spend more time with home computers, and are sent to summer computer camps. Older boys learn to work with machines, repair cars, and use a variety of tools. All these experiences develop spatial and other mathematical skills, as well as confidence in the ability to work independently.

Girls, on the other hand, learn housekeeping and nurturing skills as they play with dolls and help in the house. Of course, sewing and cooking require skill in measuring and counting, but these "feminine" accomplishments are rarely considered mathematical. Increasingly, girls are engaging in athletic activities

Barbie Doll Fears Math

Teen Talk Barbie put her foot in her mouth when she was programmed to say: "Math class is tough," one of four phrases randomly selected from a total of 270. This is just the message that little girls should not be hearing. Mattel Inc. was soon flooded with complaints from math teachers, women's organizations, and parents. Nor were the media silent. "Educators Give Barbie a Good Dressing-Down" wrote the *Wall Street Journal* (25 September 1992), ever mindful of sales volume.

Mattel's offer to exchange the offending doll for a different Barbie did not appease the opposition. Finally Mattel withdrew "Math class is tough" from the list of phrases. In a letter to the American Association of University Women, sponsor of the report describing how schools shortchange girls, the president (female) of Mattel wrote: "We didn't fully consider the potentially negative implications of this phrase, not were we aware of the findings of your organization's report." (Kevin Sullivan, "Foot-in-Mouth Barbie," *Washington Post,* 30 September 1992, A1; "Teen Talk Barbie Turns Silent on Math," *New York Times,* 21 October 1992, D4.)

like baseball, soccer, and karate, and these activities may be contributing to their improved performance in math.

A long-term study of 1,500 children from seventh grade through high school revealed that parents have a strong effect on their children's attitudes toward mathematics. In particular, mothers tended to consider math difficult for their daughters, and discouraged them from taking advanced courses.[7] These parents may have encouraged their daughters to develop a trait that psychologists call "learned helplessness," as exemplified by Harriet.

Harriet's story is told by her college math instructor. Before she even looked at a problem, Harriet was convinced that she couldn't handle it. "Frantically she sought help from anyone who would listen to her. She admitted asking more than a half dozen people for help on one problem." She finally began to make some progress when her instructor convinced her to come to her, and to no one else, for help.[8]

A study in the Montgomery County (Maryland) public schools found that many of the sexist beliefs and attitudes about girls and math still persist among parents, students, and school personnel, although not quite to the same degree as previously.[9] Similar findings from other parts of the United States and for several foreign countries were reported at the International Congresses on Mathematics Education in 1988 and 1992.

How often have mothers, more than fathers, told me: "Oh, well, I was never any good at math, either," when I discussed their daughters' difficulties in math class. A generation ago that may have been an acceptable comment. Why would they need math, when they would only marry and raise children?

In contrast to the experience of white females, black women expected to go to work. Mathematics professor Vivienne Malone-Mayes put it plainly: "Whereas many white women were simply expected to be homemakers dependent upon a providing male, black girls have been conditioned to work. Every girl expected to work hard in her life. In the fifties and sixties, education provided black women a hope that they could escape the low-paying jobs traditionally designated for them."[10]

And if these goals included the study of mathematics, why not? Men, on the other hand, could find employment in unionized industries and earn good wages without having to acquire higher education. But such jobs are fast disappearing. Today young men of color are likely to join the military as a way to acquire the education they could not otherwise afford.

It's easy for a young woman to blame Mom in the face of difficulty. When Kate (age 23), a registered nurse, enrolled in the math courses required for a degree in pharmacology, her old fear of math resurfaced. "Well, my mother's weakest subject is math, too, so I guess I got that from her."

Marion (age 40) claims that her mother was a determining factor in her career decision. "I think the message was clear from my mother that what I really needed was a line of work that would allow me to take care of a family." Although she does need mathematics in her work—as an editor she is called upon to do calculations concerning layout—she solves that problem either by ignoring it or by getting someone else to do

What Are the Odds That a Woman Will Marry?

Studies and surveys frequently are biased, proving whatever the researcher or funder may want to prove. Even for valid studies, research findings may be misinterpreted to apply more widely than the investigators had intended. The media may publicize the conclusions, particularly if they border on the sensational, without giving adequate background information. The consequences may be disastrous!

Mental health professionals reported instances of women rushing into unsuitable marriages and men abusing their wives after the media had publicized in 1986 the unpublished findings of a Yale-Harvard study called "Marriage Patterns in the United States." The study, leaked to the press by one of the authors, predicted the likelihood of marriage for single, white, college-educated women: 20 percent at age thirty, 5 percent at age thirty-five, and only one percent for forty-year-olds. For black women the probabilities of marriage were even lower. *Newsweek* and *People* went so far as to feature the marriageability story on their covers.

But a Census Bureau researcher, Jeanne Moorman, used similar data to reach entirely different conclusions: 58–66 percent likelihood for thirty-year-olds, and correspondingly higher figures for older women than those reported by the Yale-Harvard team. Her report was virtually ignored by the media, and Moorman herself was censored by her superiors.

The statistical techniques and assumptions of "Marriage Patterns" were highly suspect. Sampling was restricted to women born in the mid-1950s, and women over the age of thirty were not included in the sample, an amazing admission for a study that purports to deal with older women as well as thirty-year-olds. Divorced women and widows were excluded. Nor did the study differentiate between women who wanted to marry and those who chose not to do so, for whatever reasons.

(*Continued*)

What Are the Odds That a Woman Will Marry? (*Continued*)

The Yale-Harvard study finally appeared in a professional journal in 1989. The analysis of marriageability that had attracted so much attention was omitted completely from the publication; the researchers claimed that it was "a distraction from their central findings." (Andree Brooks, "Relationships: When Studies Mislead," *New York Times,* 29 December 1986, B14; Susan Faludi, *Backlash: The Undeclared War against American Women* [New York: Crown Publishers, 1991], 9–14.)

it. Marion's feelings about math are decidedly negative, but they have not prevented her from holding responsible positions handling large sums of money. "I'm good at figuring out what things will cost and how we will pay for them. I attribute this to my mother's having taught me to be a good money manager, rather than to any ease with figures." Evidently Marion considers "being a good money manager" and being "at ease with figures" as unrelated accomplishments!

Contrast Marion's experiences with that of a young woman today. At the age of fifteen, Stacie has already decided to follow in her father's footsteps. He is an Air Force pilot, her grandfather is an engineer, and she plans to become an astronaut. "My mother doesn't like the idea, neither do my grandmothers. My father, grandfather, and everyone else think it's a great idea."

Two children raised in the same family can have entirely different experiences. Rona (age 40) wrote that her older sister was the "brain" in the family, while little Rona was expected to be sweet and adorable. By the time Rona entered high school, she had established "a successful pattern of failure," to use her description. Fortunately for her, she encountered a geometry teacher who insisted that every student learn the concepts, and not merely memorize the material. To her great surprise, Rona found that she could actually learn math. She wrote, "It is harder *not* to understand." She began to reevaluate her self-concept. "This can be quite a threat." Now she teaches math in a junior

Women Delay Having Children and Avoid Marriage

"More women these days are reaching age thirty childless than in the past several decades. Yet, fertility levels gradually rose during the 1980s, with the number of annual births reaching their highest levels since the close of the baby boom. How could this be?"

The Census Bureau asked this question, and gave the answer. More women are having their first children after they have passed the age of thirty. In 1976 only one woman out of five who had not given birth by the age of thirty expected that she would have children in the future. By 1990 the number had risen to two out of five. During the 1980s the proportion of women who were in their thirties rose. Add on the births to women under the age of thirty. The result is an increase in the birth rate.

Another interesting bit of information from the Census Bureau concerns single motherhood. In the old days, 1960–1964 to be precise, over half of unmarried women who were pregnant with their first child had married before the child was born. But by the late 1980s, only one out of four women in that situation opted for marriage. Evidently they preferred going it alone to entering an unstable and possibly hurtful marriage. ("More Women Are Having Children Later," *Census and You* 27 [February 1992]: 7–8; "More Women Choosing Single Motherhood," Census Bureau news release CB91-329, 4 December 1991.)

high school, trains other teachers, and develops the math curriculum. She and her former geometry teacher are justifiably proud!

From childhood Carol (age 44) has always enjoyed math, but not necessarily for the obvious reasons. Her perfectionist parents imbued her with the need to be just as perfect as they. Math satisfied that need. "You can always improve on an English composition, but in math, when you're right, you're right." English prevailed, and Carol became a librarian.

Research shows that girls begin to doubt their mathematical ability about the time they enter the seventh grade. From their own families, from the media, from school, they get the message that math is for boys, that girls are not capable of doing advanced mathematics or entering careers that require a mathematics background. (In the late eighties, young women were catching up with men in the number of courses they took in high school, due in part to tougher school standards.) Rarely do girls see women as scientists or mathematicians.

This image of math as a "male domain" is so powerful as to have a devastating effect on women's confidence in their ability to do math. As the level of confidence drops, so does the level of performance. Although girls continue to earn as good grades as boys, if not better, in their math courses, they tend to believe that they lack the ability necessary to take more advanced courses. Girls who get good grades feel that they have earned them by working hard, while boys attribute their success to their superior ability. When a girl does poorly on a test, she considers herself "dumb." A boy is likely to blame the teacher or bad luck, or say that he "doesn't need any more math."

Elsie (age 53) enjoys teaching gifted children in grades four to six. She had started to "hate math with a passion" when she missed some work in the course of changing schools. "The further I went the 'behinder' I got." In college she took a remedial math class, where the professor let the students give themselves their grades. "Naturally, all of us asked for an A. Never have I felt right about that A . . . guilt seems to last."

The stereotype that "math is for boys" is held more strongly by males than by females, but it may seriously influence girls' confidence in their ability to do mathematics. This lack of confidence can affect even the brightest and highest-achieving women.

Marcia Sward is the executive director of the Mathematical Association of America and an outstanding mathematician. "[In high school] I always thought that the boys were a lot smarter than I was." She graduated summa cum laude from a women's college, where she had majored in mathematics. Graduate school was a great shock. "I suddenly discovered that mathematics was regarded as a man's subject, and I lost some of my self-confidence. . . . I wanted to get a Ph.D. But I had a long way to go to develop the self-confidence that I could do it." Of course, she did it.[11]

Luann Fears Math

Two math professors at Marymount College wrote to the Gannett Newspapers to protest Greg Evans's stereotypes about women and math in his daily comic strip, "Luann," for the week of 8–13 April 1991. The class nerd offers to help Luann with her math homework, while she confesses to feeling "stupid, frustrated, and sick to my stomach." Luann is not only poor in math, but she sees no point in learning it. Then she makes a flippant remark about the purse holding the coins in the algebra problem, and later gives up completely. In the last strip she gives the "nerd" credit for her correct solutions.

The letter concluded: "We have been striving hard to make our female students realize that they are just as good as men in subjects requiring mathematical reasoning, and that at least some of their 'math anxiety' has been caused by the popular stereotype that math is not suitable for women. Greg Evans has made our work harder." (Doris Appleby and Maryam Hastings, "Girls Stereotyped as Math-haters," Gannett Suburban Newspapers, 5 May 1991, Letters.)

Women in science and engineering courses also suffer from male sexist behavior. An instructor in an engineering institute heard a male student remark: "Those dumb girls don't know one end of a resistor from the other." When the instructor intervened, the student objected: "But professor, that wasn't a sexist remark!" In other words, he was just telling it as he saw it. These male students took it for granted that they were to conduct the experiments while the women merely took notes—carried out the role of secretary for which they were fitted by nature.[12] Is it any wonder, then, that an adolescent girl might be reluctant to be considered a "math brain" for fear that boys will shun her, or even that she might be unable to attract a husband in the future?

The attitude of male teachers can be devastating to young women's self-confidence. Although Nora (age 52), a fifth-grade

teacher, loved math in elementary school, and had "terrific teachers with high expectations and confidence in students' abilities," her high school experience was very negative. The teacher "was not interested in having girls in his math class. He also had no confidence in girls' ability in math."

Anne (age 37) did so well that she made the junior high school math team. But her interest in mathematics ended with the male high school math teacher who ignored the girls in the class, even when they raised their hands to volunteer answers. She turned to creative writing as her career.

These males reflect an important aspect of their socialization, that women's participation in and contributions to society have little value. This attitude is rarely overcome without conscious awareness and determined effort. Like the engineering student, many men don't even know that they are guilty of sexism!

A study comparing young men and women about to enter the university in Malaysia found no differences in either their attitudes to mathematics or their performance in the mathematics entry examination. In fact, on one question involving spatial relations, women did better than men. Frank Swetz explains these outcomes:

> Modern science and technology are just beginning to establish a presence in the country. Mathematics has not yet

> been culturally stereotyped as a male profession; thus, apparently there are no sex-related biases associated with it. Also in the traditional societies of Asia, Africa, and Latin America, sex-role modeling may be more conducive to familiarizing girls with mathematics and its uses than in the West. In the Third World, women are frequently small business entrepreneurs. Selling crafts and produce in the markets, they become accustomed to using weights and measures and undertaking calculations. In some of these societies, women manage the family accounts and actually own the family land. The weaving of geometric patterns, whether in rattan floor mats, baskets, or cloth, develops an appreciation for geometry in its forms. In non-Western societies, weaving is women's work and part of a young girl's training. The Malaysian girls sampled appeared comfortable in problem-solving situations involving geometric manipulations and spatial visualization. Thus, it would seem that in Malaysia, and possibly other non-Western societies, cultural and societal constraints do not discourage women from participating in mathematics and may even perhaps encourage their participation.[13]

Aldo (age 19) entered a community college in California after his education in private schools in Asia. He thinks highly of female mathematical aptitude: "My father is not good in math but my mother is quite good. Moreover, my brother and my sisters are much better than me in math courses."

Akiko, a young Japanese woman studying at a community college in Oregon, expresses a contrary view of school mathematics. "I studied math in Japan. As you know, Japanese math level is very high because we have been studying math for nine years constantly. Almost all of the students would finish until college algebra. In Japanese math system, students do not often use calculators, and cannot use them on tests. The students also are not given formulas. We have to remember all of them when we take tests. In this situation, I did not like math because I had to remember many formulas. I did not like calculation, especially easy ones. I made many careless mistakes during calculating. . . . My mother has a fear of math. Generally speaking, many women have a fear of math."

African Market Women and Math

An article in *Time* magazine extolled the perspicacity of the West African market women, who control much of the transport and trades in textiles, food, and hardware in Ghana and Nigeria. Bankers in Lagos tell of one woman who cannot write her own name, but who can get a letter of credit for 200,000 pounds ($560,000) whenever she needs one. To quote one African journalist: "They can't read or write, but they can damn well count." (Quoted from Claudia Zaslavsky, *Africa Counts: Number and Pattern in African Culture* [New York: Lawrence Hill Books, 1979], 224.)

Boys in American society are generally brought up to accept a challenge, to experiment, to take risks, and not to fear making mistakes. Girls, on the other hand, are taught to be more passive—or call it better-behaved! They are cautious about attempting new tasks. They take time to consider all the alternatives before answering questions on tests, a trait which may be a contributing factor to girls' lower scores on timed tests, in spite of class grades in math that are equal, or even superior, to those of boys.

Indeed, this cautious behavior may even help girls to earn good grades. When I was a senior in high school, I entered a contest for the trigonometry medal. Five boys and I sat for three hours after school as we tackled ten tough problems. I was sure that I hadn't won because I had solved no more than five, while the boys boasted of having done seven or eight. When the results were announced, it turned out that I was the winner with five correct solutions!

Cultural Attitudes and Mathematics

Regardless of their own level of education, African-American parents, perhaps even more than white parents of the same social class, have high aspirations for their children and look to education as the way for their children to gain upward mobility. Dolores Spikes is a fine example.

Dr. Dolores Richard Spikes majored in mathematics, and is now president of Southern University in Louisiana, the historically black institution from which she had received her bachelor's degree. Her parents "believed education to be the key to a 'better' life and career." Her father worked overtime consistently to provide for his family's educational needs. "Even among those with little formal education in my father's family, there seemed to be a natural inclination towards mathematical-related interests." Dr. Spikes passes along these words of wisdom for young people:

> Students, even those who profess "not to like" mathematics or "not to be good" at it, should not shy away from school subjects involving mathematics. I recall that *everyone* in my [segregated] high school at the time of my matriculation there took algebra and geometry. In the end, even those who managed to earn only D's were much more successful in later studies or career pursuits than students who I note today avoided such subjects prior to college or career.[14]

Less research—in fact, very little—has been done with members of minority groups than with women to determine their attitudes toward mathematics. Much of the evidence I will cite is anecdotal. Although African Americans, Native Americans, and Latinos score lower than whites in standardized tests, their scores have been rising, while those of whites have remained stationary. Young black women are more likely than black men to take advanced high school courses and to continue in college. The opposite is true of Hispanics; Latinas are more likely than Latinos to drop out of math and school. Asian Americans have the highest scores, higher than those of whites, and take the most advanced math courses. Their verbal SAT scores are far lower than their math SAT scores, probably an indication that a considerable number of Asian test takers did not grow up in the United States.[15]

Unfortunately, many young minority people don't understand how mathematics can be useful for their future jobs or college careers. For a multitude of reasons, they fail to take the

Women as the First Mathematicians

Women were the first mathematicians ever! So claims Dena Taylor in an article entitled "The Power of Menstruation" (*Mothering* 58 [Winter 1991]: 41): "The cyclical nature of menstruation has played a major role in the development of counting, mathematics, and the measuring of time. . . . Lunar markings found on prehistoric bone fragments show how early women marked their cycles and thus began to mark time."

Let's review some of the evidence. In my book *Africa Counts: Number and Pattern in African Culture,* I wrote about the Ishango bone. This incised bone, discovered in the 1950s on the shore of Lake Edward in northeastern Zaire, was originally described as a record of prime numbers and doubling (perhaps a forerunner of the ancient Egyptian system of multiplication by doubling). Alexander Marshack later concluded, on the basis of his microscopic examination, that it represented a six-month lunar calendar.

The dating of the Ishango bone has recently been reevaluated, from about 8000 B.C. to perhaps 20,000 B.C. or earlier. Similar calendar bones, dating back as much as 30,000 years, have been found in Europe. Thus far the oldest such incised bone, discovered in southern Africa and having twenty-nine incisions, goes back about 37,000 years.

Now, who but a woman keeping track of her cycles would need a lunar calendar? When I raised this question with a colleague, he suggested that early agriculturalists might have kept such records. However, he was quick to add that women were probably the first agriculturalists. They discovered cultivation while the men were out hunting. So, whichever way you look at it, women were undoubtedly the first mathematicians! (Adapted from Claudia Zaslavsky, "Women as the First Mathematicians," *Women and Mathematics Education Newsletter* 14 [Fall 1991]: 4.)

high school math courses that will prepare them for the careers of their choice.[16]

Bob, an African American, is a good example. He was persuaded to enroll in my Algebra I class in his senior year in high school, as a prerequisite for his planned career in electronics. But he was too preoccupied with weekend dates as a lead singer with a rock band to spend much time or thought on math. A few months after graduation Bob came back to visit. "You were right," he told me, as though he had made a great discovery. "I really *do* need algebra for my electronics courses!"

I had the opportunity to survey seventh- and eighth-grade students in a magnet (specialized) school for mathematics and science. Of the ninety students, thirty-nine were Latino and thirty-six African American, approximately evenly divided between boys and girls. In response to a question about future careers, many chose fields that involved some math or science, with two notable exceptions. Even in this specialized school, of the eighteen black males, five hoped to become professional athletes and one a rock star, while six of the twenty-two Latina girls chose either traditionally female or glamorous roles, such as secretary or ballerina.

It is not surprising that some African-American boys dream of becoming athletes or rock stars, the visible models of successful black males. Yale professor James Comer discusses the influences that turn young black males against academic involvement.

> One black youngster reported that he felt like a "turncoat" because he took college algebra with the college bound, mostly white, students while his buddies, mostly black, were taking applied (for daily living) math. Others have reported that it was awkward or embarrassing to be interested in math, science, or poetry, while most of their friends were more interested in the latest recording by "The Soul Singers" or "bouncing the basketball." To have academic interests in such an environment, a child must go against the tide or stand alone (or with little support) against the pressure of peers, neighbors, or the subculture. When few parents or neighbors have jobs requiring academic skills, there are few models to help develop interest or competencies in these skills even when peer pressure is not too great.[17]

Jeff Howard attributes such negative behavior to the legacy of centuries of racism, a legacy that still pervades American society. Young blacks, especially males, have internalized the stereotypes projected by the larger society and unconsciously have come to believe in their own intellectual inferiority. Howard teaches his Efficacy Model in many city schools, to "encourage young blacks to attribute their intellectual successes to ability (thereby boosting confidence) and their failures to lack of effort."[18]

The NAACP is countering the stigma on intellectual pursuits by sponsoring the Afro-American Cultural, Technological and Scientific Olympics (ACT-SO). Sixteen-year-old Brian Hooker of Atlanta was the 1986 winner of a gold medal in mathematics and a silver medal in chemistry. Hooker remarked wisely: "Students around the nation are constantly reminded of the athletic accomplishments of blacks, but those who are brilliant in the academic fields rarely get that type of recognition.[19]

Cultural factors may influence some young people to exclude mathematics from their plans. As a child growing up on the Nez Percé Indian reservation in Idaho, Edna Lee Paisano loved mathematics and excelled in the subject. But she didn't consider math appropriate as a college major. "Could I go back and say to the tribe, 'I am a mathematician'?" She felt that she had to enter a profession that would be useful to her people, and at that time she couldn't see math as useful. Her choice was sociology. Later, as the only full-time American Indian at the national office of the Census Bureau, she realized how important it was for her people to know computer programming, statistics, and demography. She designed a special questionnaire for the 1980 census which, for the first time, gave adequate information about American Indians and Alaskan Natives.[20]

Native American communities are badly in need of people trained to work in such fields as health and energy resource management. Mathematical competency holds the key to such careers.

In Latin-American families attitudes about traditional gender roles may lead to differential upbringing of girls and boys. Alberto (age 26), the pride and joy of his family, grew up in Colombia. But this very pride was his undoing. Always an A student, he was sent to a top private high school, while his sisters attended public schools. In the first semester of algebra he found the pace too fast, but was too embarrassed to admit it. He had to uphold his reputation as an A student! Alberto got through his math courses by memorizing the material. When he failed the entrance exam for a New York City public college, he was so shattered that he left. "I felt that I was retarded." Eventually, after three remedial courses, he passed the exam. He attributes his success to the tutor who insisted that he use a pencil without an eraser, so that he could analyze and learn from his errors.

Latinos in the United States, like African Americans, may have internalized subtle and not-so-subtle messages of inferiority from their teachers and the larger society. They have also been victims of inadequate education.[21] In a study of Latino (mainly Mexican-American) undergraduates at a Texas university, about one-third were found to have very low levels of confidence in their math ability. Ranking highest on the "anxiety" scale were females, older students, and prospective elementary teachers. Although one-third of the students had taken no more than Algebra I in high school and they had a poor understanding of mathematical concepts, the majority said they liked math and thought it was important for their future life.[22] These students should have had a Jaime Escalante in their lives!

The subject of the 1988 film *Stand and Deliver,* Escalante has been doing the "impossible" since 1980, teaching Advanced Placement Calculus to low-income Mexican-American students at Garfield High School in East Los Angeles. By 1987 Garfield ranked seventh in the nation in passing rates for the calculus exam. Furthermore, the majority of these calculus students are young women! Escalante expects his students to achieve at the highest level, regardless of their previous course grades and test scores, and inspires them with the desire to do well. *Ganas*—"the wish to succeed"—is his favorite word. He instills in the students a pride in their Mexican mathematical heritage, telling them that the Maya developed both the concept of zero and a symbol for zero many centuries ago. The students work hard—early morning, evening, summer. With Escalante as an example, the whole school has been turned around. Now there are few dropouts, and most students plan to go to college.[23]

Italian Americans, perhaps more than other white ethnic groups, suffer from low self-esteem, reinforced by negative, anti-intellectual stereotypes in the media and alienating experiences

The Maya, Masters of Astronomy

The ancient Maya inhabited southern Mexico and the northern regions of Central America, where their millions of descendants still live today. Two thousand years ago they had developed writing, systems of numeration, and advanced concepts of astronomy. Only in recent years has their writing been deciphered, and we are learning more about their culture as investigators translate the few documents that escaped destruction by Spanish conquistadors and missionaries, who considered them "works of the devil."

The Maya devised a positional numeration system based on groups of twenty, and were possibly first . . . to invent symbols for zero. Many stone monuments, called stelae, have dates of important events engraved upon them. These dates refer to the various calendars by which Maya priests and scribes (women as well as men) guided the agricultural year, determined the times for ceremonials, and marked the movements of the heavenly bodies.

The calendars were interrelated. Each day in the 260-day ritual calendar was identified with a number in a cycle of thirteen and a name in a cycle of twenty (like our numbered days of the month and named days of the week). Each number and each name was associated with a different deity. Meshing with this calendar was the 365-day vague year, eighteen periods of twenty named days, plus five unnamed days. The first day of the new year in both calendars coincided every fifty-two years (seventy-eight ritual years).

To mark a major event, a stela might have a set of numerals to show a Long Count, the number of days that had elapsed since the onset of the current Great Cycle, which we believe began in 3114 B.C. In one instance a date one-and-a-quarter million years in the past was calculated. Other calendars dealt accurately with the movements of the moon and the planet Venus. (Marcia Ascher, "Before the Conquest," *Mathematics Magazine* 65 [1992]: 211–218.)

in school—"for many, it is more confrontational than educational!" In New York City about one in five Italian-American students doesn't finish high school, the third-highest dropout rate after Hispanics and blacks. Their working-class families "take pride in working and bringing home a paycheck, as opposed to bringing home intangible knowledge. It is an accepted sign of adulthood," according to Douglas Blancero, director of a youth agency. But without at least a high school diploma, the jobs they seek may not be there for them.[24]

A great deal has been written about the "natural aptitude in math" of Asian Americans. The attitude of east Asian parents, both in the United States and in their home countries, is a significant factor. They have high expectations for their children and monitor their educational progress closely. They assume that all children are capable of learning. If children are not performing well, it means they just didn't try hard enough. High school students devote many hours to school work; school is their full-time job.[25]

In an article published in a high school students' newspaper, Yuh-Yng Lee tells about her feelings and experiences.

> Because I'm Chinese, my friends expect me to get high grades in math and science. When I do, they say, "Of course she got that, it's easy for her." For your information, Chinese kids are not all math and science whizzes. It's not something in our genes. I don't know about other Chinese kids, but I don't excel in math and if I get a good grade (which isn't all the time), it's because I worked hard for it.
>
> The Chinese see education as the key to a good future, and greater opportunities. That's the main reason many Chinese parents come to America. My parents had comfortable jobs in Taiwan and made a decent living. When they came here, they had to take jobs waiting tables and doing other manual work. They gave up good lives for me and I often feel guilty for not living up to their expectations. . . . All the same, sometimes when I'm home studying . . . while my friends are out partying . . . I resent my parents a lot.[26]

Remember Akiko, brought up in Japan, and her comment that many women fear math. Such remarks should help to destroy the stereotypes about innate mathematics ability.

Career Goals

"Do I need math?" The answer to that question has often been the determining factor in a high school student's choice as to whether to take more math courses once the requirement has been fulfilled. Many young women do not have clearly defined career goals when they are in high school. As a result, they fail to take courses that turn out to be prerequisites for their college majors and future careers. Those who do have definite plans generally orient their goals toward helping humanity and working with people. They make the assumption—mistakenly—that math is not necessary for these careers.

Mathematics is called the "critical filter" that excludes people, especially women, from entering many professions. The term is due to Lucy Sells, a California sociologist. In 1972 she surveyed the entering class at the University of California at Berkeley, and was shocked to find that 57 percent of the males but only 8 percent of the females had taken enough high school mathematics to qualify for most of the forty-four majors. Without this math background, these young women were restricted to such traditionally female, low-paying fields as elementary education, guidance, and the humanities.[27]

Lucy Sells's specialty is the analysis of female and minority participation in mathematics and science courses. Her autobiography reveals that she experienced many of the negative factors we have discussed—confusion resulting from missed classes in elementary school, an extremely sexist male teacher in high school, a guidance counselor who told her she could drop math because she would never need it, and no clear career goals in college. Eventually she drifted into the graduate sociology department. "I learned quickly of the dread statistics requirement, which was the brick wall for doctoral candidates in the social sciences and the school of education."

But Dr. Sells did not go under. The concluding paragraph of her autobiography is truly inspirational:

> I had a great professor, an even greater teaching assistant, with incredible patience, who helped me discover that statistics was *fun*! I worked *very* hard, attended all of the volunteer lab hours, and got a D on the first midterm. I salvaged an A in the course, and got A's throughout the required advanced quantitative methods course that had kept eight or ten otherwise brilliant grad students from taking their Orals exams. I went into the final knowing that a Teaching Assistantship rested on cooling the exam. For the first two hours of the three hour exam, my mind froze, and I couldn't make sense of the problems. I finally went to the rest room, got a drink of water, and sat down and did the exam in the last hour, getting an A− *and* the Teaching Assistantship. For me, the greatest joy of teaching statistics was to see puzzled faces light up with understanding when they finally got it. I wish that joy for every teacher of things quantitative, as well as for every student."[28]

Not only are women led to believe that math is "hard," but they also associate it with the "hard sciences" like physics, and with the applications of math and science in our society. Ruth (age about 50) gives her impression of math and science as it is practiced: "What I do know and can attest to is that mathematics (and science) in this culture is undoubtedly perceived as a function of the war machine, and it is bloody hard to convince a smart girl or woman to *go to war.*"[29]

In a four-year study of "unlikely achievers from disadvantaged backgrounds," Charles C. Harrington of Teachers College, Columbia University, identified the traits that these successful people had in common. Foremost was internal control, the ability to see themselves as in charge of their lives. This trait is strongest among successful black women, who had to overcome both racism and sexism to achieve their goals.[30] Here are the stories (in brief) of several women of color who use mathematics and had to overcome tremendous hurdles to fulfill their aspirations.

Eleanor Jones, now a professor of mathematics at Norfolk State University, was prevented by the segregation laws of the state of Virginia from pursuing doctoral studies in that state. She was, however, granted tuition and travel costs to study in the North. While supporting her two small children, she earned her doctoral degree in mathematics at Syracuse University in 1966.[31]

Many among my respondents displayed this trait of internal control. Daniela (age 21), a mathematics major, was about to graduate from a historically black university at the time she completed her math autobiography. Accelerated in high school mathematics, she took Algebra II in tenth grade. "I was the only black in my class. It made learning the material a job, since I talked to no one in the class." By twelfth grade she was taking a college-level course. "Everyone made math seem so complex and the material insurmountable!" she said.

As a girl growing up in Puerto Rico, Carmen Gautier loved math and science. But such interests were not for Puerto Rican girls in the 1940s. "I used to ask my teachers if I could take trigonometry. They told me girls don't need to take that. I was advanced in chemistry, but they told me girls can't take physics." Undaunted, she went on to earn a doctorate in political economy, a field requiring a considerable knowledge of mathematics. Now she is director of the Social Science Research Center of the University of Puerto Rico, teaches at the university, and writes in defense of Puerto Rico's land, people, and resources.[32]

The early life of Helen Nealy Cheek, a Chocktaw (Native American) mathematics educator, gave little indication of her future success. A high school dropout, she was a divorced mother of two children at the age of 21. She returned to high school and continued her education while working and teaching elementary school on an Indian reservation and elsewhere. At the age of 43 she earned her doctorate in elementary and Indian education at Oklahoma State University. The author of many articles, recipient of awards, leader of numerous professional organizations, she was particularly interested in equality for women and Native Americans.

In her teaching she demonstrated the interrelationships between patterns in life and patterns in mathematics, using design in beadwork, basketry, and patchwork to illustrate mathematical concepts. She taught mathematically based Native games to children while she engaged their parents and grandparents in various mathematical activities, a form of Family Math. Her death from cancer in 1985 at the age of 48 was a keenly felt loss.[33]

Asians, whether born abroad or in the United States, are most likely of all groups to take advanced math and choose quantitative majors in college. According to researcher Sau-Lim Tsang, "There is no research support for the hypothesis that Asian Americans have a high level of innate mathematical ability." Sociological factors play a significant role. Many Asians coming to this country are well-established engineers and scientists, while others came as students and decided to remain. They make a point of investing in their own and their children's education, to offset discrimination and to obtain upward mobility. Recent immigrants choose to study mathematics and science, subjects that don't demand proficiency in English.[34] And the jobs are there for the asking. As young electrical engineer Hai Van Vu put it: "I came here ten years ago from a small village in Vietnam. . . . I had a problem with language—but not with mathematics! The symbolic notations of math were a universal language I could understand as well as anyone. I could compete! Patience and concentration in math and science gave me entry into a fascinating field. . . . I knew that after four years of college, if I did well, I was almost guaranteed a job—a *good* job."[35]

The Computer Scene

Technology is rapidly changing both the educational scene and the workplace. The gap between the "have-computers" and the "have-nots" results in some people being left out of the picture. Those people are generally women, people of color, and working class.

Men tend to take over in a learning situation that involves both sexes. At a computer workshop for elementary teachers, I introduced the basic aspects of computer programming, using as an example the game of tic-tac-toe. The group consisted of ten women and one man, but that one man, obviously experienced with computers, dominated the scene, calling out answers to all the questions I posed before the women had a chance to think. Finally I had to remind him that everyone deserved the opportunity to participate.

If a grown man, a teacher at that, can't resist the chance to show how smart he is, then imagine the situation in an elementary classroom. The school has just acquired several computers. Although some teachers were able to enroll in a course organized by the school district, there was no space for Ms. Brown this semester. She hopes to take the course next term, but meanwhile the computer sits in her classroom. She asks who has a computer at home and can show the class how it works. Jimmy volunteers. Naturally, the boys flock to watch Jimmy's demonstration and, perhaps, to share their own experiences, while the girls feel left out. And so begins a new round of fear and anxiety.

Boys, more than girls, have home computers and work with them, go to computer camps and after-school programs, play video games (girls tend to reject them as too violent and competitive), and look forward to careers involving computers. To make matters worse, computer ads have usually addressed their message to boys.

This may be changing as computer companies realize that women, too, are capable. Now it's the older people who are maligned. According to an item in the *Bulletin* of the American Association of Retired Persons (AARP), an ad showing a young woman at a keyboard has her saying about her mother: "She still thinks software is a nightgown." The author went on to discuss the involvement of older people with the computer culture. As just one example, the University of Miami is offering a course called: "How to talk computers with your grandkids."[36]

It was a woman who originated the whole field of computer programming—Lady Ada Byron Lovelace (1815–1852), daughter of the poet Lord Byron and amateur mathematician Annabella Millbanke. In 1834 Charles Babbage designed a machine he called the "analytical engine," a forerunner of the modern computer in spirit, although it was never built. Ada, Countess of Lovelace, translated from French into English a paper describing the functioning of the analytical engine. In the course of the translation she added so much new material that she is now credited as the world's first computer programmer.

Ada Lovelace signed her work with only her initials—women at that time did not write technical papers! For nearly ninety years after her death, her work was completely neglected. In fact, the first computers, produced in the early 1940s, would have been far more efficient if the principles she set down had been followed. Her ideas were rediscovered around 1950 by several mathematicians. The computers developed at that time used punched cards, just as Lady Lovelace had described.

Punched cards are now a thing of the past, but Lady Lovelace's name lives on in the advanced computer language called ADA.[37]

The business world is profiting from women's inventiveness. In the late 1950s Rear Admiral Grace Hopper originated a system that permitted the programmer to feed into the computer

English words instead of symbols, thus making the machine more user friendly. Subsequently her ideas were developed into COBOL, the most widely used computer business language, by a team in which Jean Sammet and other women played a prominent role. Sammet was also heavily involved in the development of the language ADA.[38]

Everyone—young, old, male, female, of all racial/ethnic backgrounds, can learn this vital technology. All they need is the opportunity!

CHAPTER FIVE

Our Schools Are Found Wanting

It was characteristic of my parochial education to be discouraged—actively discouraged—from questioning. Starting with religion first thing in the morning, we were to accept what we were told. How did this affect math learning? There was one right way to do things; one right way to think. . . .

[In college] I remember being in an anthropology course I'd actively sought to initiate and organize, along with two other classmates. We wanted a course on critical thinking, developing thinking skills. The professor who agreed to teach the course began the first class with a problem, the response to which decided for him who were the A students and who were all the rest. The problem turned out to be an Algebra word problem, and familiar to those who'd had Algebra. Major setback in my esteem and confidence. Dissatisfaction at this deficiency planted another seed. I wanted to learn and stop feeling so inadequate mathematically.

—Mary Jo Cittadino, Mathematics Educator, EQUALS, University of California at Berkeley

Maria was a college student with an all-consuming, burning desire to teach. As a Puertorriqueña, she knew how it felt to be a Spanish-speaking child in a country where everyone was expected to speak English. To help youngsters who were in the same situation, she organized classes in her church to tutor children in reading and other school work.

Imagine Maria's frustration when she learned that only 16 percent of her answers were correct in the sixth-grade-level arithmetic test administered as part of my college course on teaching elementary school mathematics. A resident of that devastated section of New York City known as the South Bronx, she had failed first-year algebra in the local high school. When she failed a second time, her teacher said, "You're a nice girl, Maria, and I know that you try hard. I'll pass you, but promise not to take any more math." No doubt he thought he was helping Maria by granting her "social promotion." It turned out to be no favor.

Fortunately for Maria and other students in her predicament, the college had just opened a tutoring laboratory. By the end of the semester she had raised her score on the required arithmetic test to 50 percent. Although she had tripled her original grade, she passed the course with a deficiency in arithmetic. Undaunted, she continued to develop her skills until she had mastered the subject. If only she had received help years before, she might have been spared the terrible disappointment of failure.

Maria represents a success story. She went on to become one of the all too few Hispanic teachers in the United States. A national study found that 45 percent of Mexican-American and Puerto Rican youths never complete high school, although their aspirations are as high as those of other groups.[1] Overall the dropout—some critics say pushout—rates are more than three times as high for poor children of all groups as for affluent children.[2]

Maria's experiences with school mathematics exemplify the faults described in a 1983 report by the National Science Board Commission on Precollege Education in Mathematics, Science and Technology. It stated that although "virtually every child can develop an understanding of mathematics, science and technology if appropriately and skillfully introduced at the elementary, middle and secondary levels, . . . substantial portions of our population still suffer from the consequences of racial, social and economic discrimination, compounded by watered standards, 'social promotion,' poor guidance and token efforts."[3]

Mark (age 23) works as a receiving clerk while attending college. He relates his high school experience: "I took three math courses, failed one, and received straight D's through the rest of them. Of course, the school just moved me on ahead anyway, regardless of my grades. I now realize that my skills are lacking in mathematics, and I am enrolled in a review math course at the university." Mark knows that "social promotion," in the long run, only held him back.

In a dramatic speech during her tenure as Hunter College president, Donna Shalala called for fundamental, systemic reform of the entire New York City public school system. In colorful language, often quoted since, she declared: "I believe the barrel is rotten, not the apples. We have allowed a rigid, inflexible, dehumanizing educational system to develop, a system that works successfully too few times. It is that system I call the rotten barrel. The apples, of course, are the children, the teachers and the administrators."[4] Although she was speaking about New York City, her condemnation might apply to many large city systems.[5]

For a select few, New York City schools are superb. In 1988, 122 of the 300 semifinalists in the annual Westinghouse Science Talent Search were seniors in New York public schools, over half from two selective schools from which have come Nobel laureates and top-level professionals in science, mathematics, and other fields. Several local communities have established magnet schools and alternative schools that serve their students well.[6]

Our educational system is a reflection of our society. We produce a well-trained scientific elite comprising mainly white middle- and upper-class males, while the rest of the population is left behind.

In the past, few voices were heard on issues of educational equity. One such voice was that of the noted African-American scholar Doxey Wilkerson. In a 1939 report he claimed that the differences in scholastic achievement between blacks and whites resulted from schooling, not from any inherent biological defect in African Americans. He called upon the federal government to "reverse the process and become an instrument of positive educational changes that would correct basic inequities within the society."[7]

Few heeded his words at the time. Blacks, for the most part, were relegated to low-paying jobs requiring little education. Today the situation is different. For many and varied reasons, our society has become aroused. The slogan now is "equity and excellence."

Why has the educational system failed to serve so many people? Why do most of our children rank so low in mathematical achievement, compared with children in other countries?

Let's look at our schools, with their inadequate funding, segregation, discriminatory practices, and poorly trained teachers. How these factors affect fear and avoidance of mathematics should become clear as we discuss the inequities and shortcomings of our educational establishment.

Schools: Segregation and Funding

When you see a newspaper article about segregated schools, it usually refers to schools in which children of color are the majority of the student body. Equally segregated are those schools, generally located in the wealthier suburbs, that serve primarily white children. Both populations are shortchanged, but in different ways. In most large city public school systems, so-called "minorities" constitute the majority of the students. Almost half of African Americans in the North attend all-"minority" schools, compared with only one quarter of black students in the South, the part of the country where the schools are now the most integrated. Segregation of Hispanic students has been increasing in recent years. Even in those districts that have taken steps toward desegregation, segregated local schools and tracks within schools result in schooling that is separate and unequal.[8] Students in many rural districts and small towns are also victims of segregation.[9]

A school that serves children of color may be of the highest quality. Garfield High in East Los Angeles, discussed in the previous chapter, is a shining example of excellence. Another example is Dunbar High School in Washington, D.C., established in 1870 as the first public high school for blacks in the United States. Until it became a neighborhood school in the 1950s, Dunbar attracted the best and brightest black students from all parts of the country. Among its illustrious alumni were Dr. Charles Drew, who devised the method for storing blood plasma, and Dr. Evelyn Boyd Granville, one of the first two African-American women to earn a doctorate in pure mathematics.[10]

A current example is A. Philip Randolph Campus High School, on the grounds of the City College of New York. Mostly black and Latino, entrants are chosen at random by computer from thousands of applicants. Half of the students come from poor families. Not only do they take eight academic classes every day, but some choose extra sessions in specialties such as engineering and medicine, in addition to mandatory math and science. Almost every graduate goes on to college.[11]

But these are exceptions. For the most part, schools in which the student body is composed mainly of people of color are inferior schools, through no fault of the students or their families.

The brilliant playwright Lorraine Hansberry told of the substandard education she received in a segregated Chicago elementary school:

> One result . . . is that to this day I cannot count properly. . . . To this day I do not add, subtract, or multiply with ease. Our teachers, devoted and indifferent alike, had to sacrifice something to make the system work at all, and in my case it was arithmetic which got put aside most often. Thus, the mind which was able to grasp university-level reading materials in the sixth and seventh grades had not been sufficiently exposed to elementary arithmetic to make even simple change in a grocery store. This is what we mean when we speak of the scars, the marks that the ghettoized child carries through life.[12]

In some cities, such as Louisville, Kentucky, and Charlotte, North Carolina, where court-ordered busing is strongly enforced, school integration is working and has even led to residential integration. In northern cities, on the other hand, where housing segregation leads to school segregation, there is little or no effort to desegregate the schools. In the Chicago, Boston, and New York public schools, only about one child in five is white, although the proportion of whites in the general population is much higher. Many white parents choose to send their children to private or parochial schools.[13]

A reader of the 3 June 1987 issue of *Education Week* would have been shocked by a photograph of a class held in a washroom of a Bronx (New York) school, with a toilet clearly visible. Not only are the New York City schools overcrowded, but the existing buildings are in a terrible state of disrepair. Robert Wagner, who was then the Board of Education president, commented, in reference to the deplorable condition of a high school formerly attended by whites, "Now it is Hispanic and black, and who cares? What a signal that sends."[14]

New York City is just one example of a situation that afflicts many large cities, and crumbling buildings are just one symptom of general deterioration.

The Penalty for Segregation

In 1985 a Federal judge ruled that Yonkers, New York, had intentionally discriminated against African Americans in housing and education. The city agreed to end school segregation, but the City Council refused to go along with the court order to build integrated low- and middle-cost housing in predominantly white areas. In 1988 Judge Leonard B. Sand declared that the city was to be fined for every day that it refused to comply with the order. The fine of $100 was set for August 2, 1988, and doubled every day thereafter.

Here is the sum that failure to comply would cost the city of Yonkers:

Day	Daily fine	Cumulative fines
1	$100	$100
2	200	300
3	400	700
4	800	1,500
5	1,600	3,100
6	3,200	6,300
7	6,400	12,700
.	.	.
.	.	.
.	.	.
22	209,715,200	419,430,300
.	.	.
.	.	.

As you can see, the fines mounted at an astronomical rate. By the twenty-second day the total sum would have exceeded the city's annual budget of $337 million. However, on the seventh day the U.S. Court of Appeals suspended the sentence, and later ruled that the daily fine was not to exceed one million dollars per day. With this threat hanging over them, the four recalcitrant City Council members gave in and agreed to a desegregation plan. ("Yonkers: The Cost of Defiance," *New York Times*, 27 August 1988, 28.)

When I was thinking about writing this book, I started to collect articles about funding for education. Jonathan Kozol's powerful 1991 book, *Savage Inequalities: Children in America's Schools,* made my collection superfluous. The book opens with a look at the devastated schools of East St. Louis, Illinois, the most distressed small city in the country, where raw sewage flows into basements and fumes from local chemical plants poison the air. These same chemical companies have incorporated their own town so that they do not have to pay taxes to support the schools of East St. Louis, a city whose population is almost 100 percent black. In contrast, a high school in the predominantly white suburban town of Winnetka, Illinois, sits on twenty-seven acres and offers a wealth of courses, computers, guidance services, and athletics.

Kozol writes about the Bronx, New York, school located in a former skating rink on a business street, where an elevated train runs past every few minutes. The building has no windows, and there is no sign on the outside to indicate that it houses a school. Several years ago I gave an in-service course at that school on the subject of "Multicultural Mathematics Education." An ethnically diverse group of Bronx teachers sat in a stuffy room for two hours after hectic days with their own classes. They turned out to be the most creative and enthusiastic class I have ever worked with!

Commentator Anna Quindlen also visited this windowless Bronx school. In her column, "Without Windows" (*New York Times,* 16 December 1992), her indignation is clear. "You learn your worth from the way you are treated," she writes. "The children at P.S. 291 learn that they barely deserve oxygen. . . . And yet we assume they will magically develop a sense of the dignity of humanity. When they are grown-up disappointments, in trouble or on welfare, we wonder what went wrong. . . . We have a system that discriminates against the poor in everything from class size to curriculum. Universal school is a myth, a construct of our egalitarian imagination."

Kozol finds that schools are now more separate and more unequal than they were twenty-five years ago. Funding from local sources is based on property taxes. A community with a wealthy tax base can afford funding for the kind of education that all students deserve. As a result, a suburban district may spend five to ten times as much per student as a nearby city school. Attempts to equalize spending among districts within a state are based on the "Robin Hood" approach—take from the wealthy to give to the poor, a solution that rich districts resist with all their might and power.[15]

Meanwhile, the federal share of funding for education fell from 9 percent to 6 percent during the 1980s. According to an article in *The Nation,* "Among the top nineteen industrial nations, the United States ranks seventeenth in public spending for education, and dead last in compensation of teachers. The United States is also the only major industrial nation to promote a radically inequitable school-financing system based on local property taxes."[16]

How does the shortage of funds affect the teaching of mathematics? Programs for remedial math help are usually not a top priority, and most schools will forgo them when funds are not available. The same is true for staff training in the pedagogy of mathematics education. Low salaries and abysmal conditions make it difficult for poor districts to attract well-trained teachers.

As Kozol noted in his survey of dozens of neighborhoods, most schools now have some computers, but many students are not reaping the benefits of the new technology. As with other resources, computers tend to be concentrated in affluent schools, while African-American students and children with limited English proficiency have the least access. Moreover, poor schools can't afford to train teachers in the appropriate use of computers.

An extreme example of inadequate funding is that of Toledo, Ohio. The city was forced to close its schools several times during the 1970s because it ran short of money for education. With no other way available to express their resistance to high taxes, the voters took out their resentment on the kids by voting down one school budget after another. When new math programs were introduced, the city had no money to educate teachers in the use of the materials.[17]

Inside the Schools: Tracking, Testing, and Classroom Environment

A look inside the schools yields futher insight into the conditions that bring about fear and avoidance of mathematics. Let's examine the school practice known as tracking, often called "sorting"

by its critics. In many schools, children entering the first grade have already been tracked, on the recommendations of their kindergarten teachers or on the basis of a child's performance on a standardized test, into high-, average-, and low-ability classes. Later, a child's score on standardized reading tests generally determines class placement. No matter that the child is a math whiz; if the reading score is low, perhaps because the child is not a native speaker of English, the child goes into the low-ability class.

"I like to take the lowest-level class," remarked my friend Paula, a first-grade teacher in a predominantly Hispanic school. "Nothing is expected of them, and so I can do as I want." And she has achieved outstanding results with some of her pupils.

The label "slow learner" stigmatizes children immediately. They get the message that little is expected of them, and they perform accordingly. Once they are placed in the slow or remedial track, there is little possibility that they can take academic mathematics courses in high school, as Tami (age 17) discovered the hard way.

> [My fear] began in the ninth grade when I had pre-algebra. That was my first time ever having a class other than general [lowest level] math and everything looked foreign to me and I didn't understand the class. I flunked one semester and had to go to summer school. In tenth grade I went back to general math because I hated algebra. Then in eleventh I decided to take algebra because I wanted to go to college. I never was good in math . . . I barely passed. I never to this day have overcome my fear or dislike of math.

In a comprehensive study of New York City high schools, Lisa Syron found that 35 percent of the students in the senior classes of

1985 and 1986 had never completed a high school level mathematics course, such as first-year algebra. In schools that were predominantly (99.9%) minority and also low-income, three out of five female seniors did not complete first-year algebra or its equivalent, and only 6 percent, or one out of sixteen, had taken three or more years of academic math. To fulfill the two-year math requirement, these students were placed in purely remedial courses or in "Record Keeping." Often, attendance was used as the measure of a student's effort.[18]

These policies hurt students in several ways: they limited their access to opportunities requiring mathematical competency, they tolerated learning climates that discouraged achievement, they imparted the message that the students were incapable of learning, and they allowed a situation to continue in which students were bound to lose confidence in their ability to do math.

Some students do brilliantly in spite of unfavorable conditions. Kathy, a twenty-year-old black Latina, survived five predominantly minority, low-income New York schools to become a college math major. All through school she had a "passion for math." She concluded her math autobiography with the emphatic statement: "I am dedicated to the Art of Math!"

Jim (age 17) is another young person who was determined to learn in spite of the obstacles. When he entered high school in the tenth grade, he asked to take geometry. Although he had completed the prerequisites in junior high school, his grades were not considered good enough for academic math, and he was tracked into a low-ability class. Determined to attend college and to become a drug counselor so that he could help youngsters on the street, Jim continued his battle with the school authorities. He finally was allowed to take geometry in his senior year and did well in the course.

An important two-year study of schools in wealthy Montgomery County, Maryland, points out the consequences of tracking. In the years preceding the study, the county had instituted a variety of practices to improve the mathematics performance of female, black, and Hispanic students, among them innovative curricula and staff sensitivity training. Although the performance of all these groups improved during those years, and both test scores and participation rates were well above the national averages, the differences in achievement among the groups persisted.

Female students took somewhat less high school math than males, received higher grades in class, and did as well as males on standardized tests, except for the Scholastic Aptitude Tests. Both male and female students scored about fifty-five points above the national average on the SATs, but the gender gap of about fifty points persisted.

White and Asian students continued to outperform black and Hispanic students. These differences were observed as early as the first and second grades, and were confirmed by standardized test scores in the third grade. Once a student fell below the standard for his or her grade level, that child was not likely ever to catch up. These differences in progress grew greater as children advanced through the grades. By the time they reached high school, some students were no longer eligible to take advanced math classes. Sixty percent of the black students and 52 percent of the Hispanic students left high school having completed at most geometry or a lower-level course, compared with 32 percent of whites and only 13 percent of Asians. Later in the chapter we will explore factors that were found to lead to these differences, in particular teacher expectations and the differentiated math curriculum.[19]

A significant study analyzed math enrollments in six predominantly or entirely African-American high schools in an eastern city. The families of the students in the six schools had similar income and educational levels, yet the students differed markedly in several respects from school to school. These were: proportion of students in academic versus nonacademic math courses, the percentages of male and female students taking math, and the numbers of students taking math in their senior year. The authors conclude: "Apparently factors in the school environment—the kind of counseling, the amount of teacher-student interaction, and the leadership shown by administrators—make a substantial difference to minority students."[20]

These factors made a difference to Karen (age 21), a college student and radio broadcaster. Discussing her experiences in California schools, she wrote: "I don't think I had any real fear of math. I believe my high school teachers and counselors instilled in me a feeling that I couldn't comprehend or compete with my peers in grade-level course work. As a result I had always been placed in lower-level classes all through my high school career.

Now I know that is not true and I will soon be able to overcome my mathematical inadequacies."

Students with limited English proficiency often are deprived of the kind of instruction that would enable them to achieve their potential. Investigation by the U.S. Civil Rights Commission into the experiences of Mexican-American students in the Southwest revealed numerous instances of neglect and abuse on the part of school personnel. These included inadequate facilities, unwarranted placement in classes for retarded children and in nonacademic tracks, and even prohibition of the Spanish language on school grounds. The courts have outlawed some of these practices, but it is more difficult to change the attitudes of school personnel.[21]

Hispanic women are especially hard hit by such practices. In 1981 only two out of five had completed high school. In high school they are tracked into such vocational courses as home economics and clerical work, leading to typically female and low-paying jobs. Role models are few, and teachers tend to have negative views of their ability. As one Latina put it: "You know the thing I hated most about school was those teachers who acted like you were stupid just because you were Hispanic."[22]

Andrea, from Guatemala, entered a New York City high school speaking almost no English. She relates her experiences with math:

> Now I am weak in Math, but I did well in Math in Guatemala. When I came here, I passed the first course because I was familiar with the material. I failed the second course because I did not understand the language, and I had a conflict with the teacher. They did not make me take the course again. Instead, they gave me a consumer math class. I was very upset. They should have allowed me to take the class again. . . . At that time I was so frustrated and angry that I just took the consumer class so I could graduate. . . . The standard of education in Guatemala was higher. . . . I was ashamed of being Hispanic and of the way I spoke. . . . This society is ethnocentric. People don't care about what happens outside of it.[23]

Do children gain by placement in different ability-level groups? There is not a shred of evidence to indicate that any but the high-level children benefit from such placement. As a matter of fact, all the evidence points in the opposite direction. The average and low-level math curriculum is boring. Children are deprived of positive interaction with their peers while they work on routine exercises in their workbooks. Teachers of these classes tend to emphasize discipline and conformity rather than the development of thinking skills. Years ago this type of education was considered adequate preparation for unskilled factory jobs. Today's students must be prepared for higher goals.[24]

Many educators urge that tracking be abolished altogether in favor of classroom organization based on cooperative learning groups. A model for such a program is Finding Out/*Descubrimiento* (FO/D). Several hundred children in grades two, three, and four, predominantly Mexican American with varying degrees of proficiency in both English and Spanish, participated in the program. A similar number of children in a bilingual program served as a comparison group. Both groups had the same curriculum except for the FO/D activities, one hour a day for fourteen weeks.

The FO/D program consisted of approximately 150 math and science activities involving measuring, counting, estimating, grouping, hypothesizing, analyzing, and reporting results. Students worked in small mixed-ability groups and were encouraged to discuss their work among themselves in English or in Spanish, as they preferred. All the materials were presented in cartoon format with texts in both Spanish and English. The children helped one another to complete worksheets in either language.

The children's progress was observed and measured in many ways during and after the fourteen-week period: intellectual development; math, science, and reading achievement; proficiency in Spanish and English; individual and group behavior. Although both the FO/D and comparison groups gained in math computation, a skill generally taught by rote memorization methods, the FO/D children far outdistanced the others in mathematical concepts and applications, as measured by standardized tests. Gains in all other areas were also significant. The researchers concluded that the most important factor influencing learning was talking and working together in small groups, which was far more effective than consulting with teachers, particularly when teachers are not immediately available. These kids were motivated to use their time productively in carrying out their assignments because the activities were challenging and the environment was social. Children improved, too, in their use of written language and their level of accuracy. The outcomes demonstrated the value of programs based on cooperative learning groups engaged in interesting activities, the direct opposite of the usual bilingual curriculum with its emphasis on rote memory skills and individual seat work.[25]

I have referred often to scores on standardized mathematics tests. They are used to compare groups of students in different programs; to compare classes, schools, and systems; to determine how individuals are to be tracked; to make decisions about promotion or flunking. How valid are these tests? Not valid at all for assessing individual students, say their critics, and perhaps not even for comparing schools and systems. Apart from outright fraud, these multiple-choice tests are vulnerable on several grounds. A multiple-choice test can measure only a narrow range of skills. It cannot measure creativity or problem-solving aptitude, abilities that are valuable in real-life situations, as well as in advanced math courses.

Yet teachers are forced to devote hundreds of hours of class time each year to teach for the tests. The tests determine the curriculum to such an extent that school districts hire experts in curriculum alignment to adjust the texts and lessons to the test items.

The brunt of this mania for testing is borne by low-income and minority children. Their schools are under the greatest pressure to raise the scores. Federally funded programs for these children demand accountability, and achievement is measured by test scores. But as we have seen, programs such as FO/D prove their success even when measured by the scores on standardized tests; children gain more by engaging in interesting activities than by repeating memorized facts. It's time to devise better ways to assess learning. As a matter of fact, new types of assessment are already under consideration.[26]

Teachers, the Key Factor

A common thread runs through the math autobiographies I received from African-American students majoring in mathematics, science, and engineering at a predominantly black

The Lake Wobegon Effect: Can All Children Be Above Average?

Is it possible for most children to have "above average" scores on standardized achievement tests? "Average" on such tests is usually defined in terms of grade level. Half the students in that grade have scores at or above this point, while the other half have scores at or below this point. Students are ranked as being at grade level, or a certain number of years above or below their grade level.

In 1988 a West Virginia physician, Dr. John J. Cannell, filed consumer-fraud complaints against four major test publishers after he had learned that in all states that mandated state-wide testing, children were achieving "above average"; each state's average scores were above the grade levels set by the testing organizations. His report was dubbed "the Lake Wobegon report," in reference to Garrison Keillor's mythical community in which "all the women are strong, all the men are good-looking, and all the children are above average."

The test makers contended that the definition of "average" had been established six or seven years earlier, and that children had generally improved in the intervening years. Critics of these testing programs gave other reasons for the rising scores, among them the practices of adjusting the curriculum to the narrow content of the tests and devoting much classroom time to drilling for the tests.

It would seem that educators and test makers are not above misleading the public as to how well children are learning. (Daniel Koretz, "Arriving in Lake Wobegon: Are Standardized Tests Exaggerating Achievement and Distorting Instruction?" *American Educator* 12 [Summer 1988]: 8–15, 46–52.)

university. Somewhere along the way they had encountered good teachers, people who were able to make math interesting and enjoyable and who encouraged them to do well. Here are some excerpts:

- *Mona (age 19, engineering major):* "In high school my love for math increased every year because I had excellent teachers who encouraged me to move on."
- *Jonathan (age 20):* "At high school I had excellent teachers who really were concerned about the students excelling beyond high school into college. I really learned the material covered in math, and I thank them even today."
- *Rhoda (age 23, chemistry major and practical nurse):* "I believe that the most important thing that has helped me to overcome my fear of math . . . was the caring demonstrated by my instructors. I think that if more teachers realized the effect that they could have on students, they would use their abilities to help produce more positive-thinking math students."
- *Matt (age 18):* "Not until my second year in high school did I show any interest in mathematics—probably because my teacher really made class interesting; from then on I tried to learn the techniques of math."

Dr. Helene Walker belongs to an older generation. She had attended a coed high school with a female math department chairperson. "There was no gender differentiation. The inspiration of my high school teachers, especially the math team coach (a man), encouraged me to major in mathematics in college."[27]

Dr. Walker was more fortunate than many. From preschool on up through graduate school, teachers and counselors are guilty of sexist behavior. In the early 1980s Drs. Myra and David Sadker conducted an extensive study in school classrooms to compare the attitudes of teachers toward girls and boys. Although the majority of elementary school teachers are women, the Sadkers found that they devoted more time and attention to boys than to girls. Particularly in math and science, teachers are far more apt to call upon boys, expecting that they, rather than girls, will give the right answers. When a girl is called upon, she is given less time to respond before the teacher turns to another student. Even feminist teachers, those who set out to be fair, are

The Teacher As Inspiration and Role Model

Fern Hunt is one of three contemporary mathematicians profiled in the "Black Achievers in Science" traveling exhibition organized by the Chicago Museum of Science and Industry. A professor at Howard University, she didn't really like math in elementary school. Inspiration came from a black junior high school science teacher. He not only made science fascinating but also urged her to apply for a science honors program at Columbia University. Eventually this interest led to a doctorate in mathematics at New York University. Besides teaching mathematics, she has devoted time to doing mathematical modeling in cancer research at the National Institutes of Health.

Fern Hunt's path was not easy. "To be honest, when I was going to college and in graduate school I don't think I got a lot of encouragement. By and large I had to seek out the professors and didn't get a lot of good advice about what to do and when to do it. In graduate school I think the professors probably thought it a bit odd that there was a Black woman studying mathematics at that level. I was very conscious of having to validate my own legitimacy."

She is optimistic about women's involvement in mathematics. "There has been a tremendous change in women's consciousness. . . . There are going to be little girls coming along with mathematical talent and with a desire to pursue mathematics." (Harriet Jackson Scarupa, "Women and Mathematics," *New Directions* [July/October 1983]: 17–23.)

guilty of sexist treatment, as revealed by videotapes of their classrooms in action.

Boys are praised for logical and independent thinking, while girls are rewarded for neatness, good behavior, and sensitivity to others. Attempts on the part of girls to break away from these stereotypes may be disastrous. A follow-up study on poor children in an enriched early childhood program found that by the upper grades the boys had generally achieved more than their peers, but the girls had not excelled. On the contrary, when they showed curiosity and independent thinking, they were put down for "sassiness."[28]

Most of the teachers in the Sadkers' study were unconscious of their bias. As an example (not in the Sadkers' study), a male teacher, meaning no harm, made this devastating comment to a boy who was slow in solving problems: "You should sit on the girls' side of the room!" To the teacher's credit, he was willing to change his behavior when his errors were revealed to him.[29]

In a study of female participation in advanced high school math and science courses, Patricia Casserly found that teachers were most influential in recruiting young women to these classes. On the other hand, guidance counselors, who are in a position to play a crucial role in students' futures, are themselves ignorant of the mathematics requirements for many careers, and are often downright discouraging. Concerned about college admissions and imbued with stereotypical ideas about girls' lack of math aptitude, they convince young women to take courses in which they are likely to earn high grades. "Why spoil your record with a hard math course?"[30]

When Ethel (age 38) did poorly on an eighth-grade math test, her counselor advised against enrolling in algebra. Fortunately Ethel ignored the advice. The outcome—an A in algebra! But this positive experience did not help her to overcome the conviction that she had no aptitude for more advanced math.

This bias on the part of guidance counselors may be found even in those who consider themselves confirmed feminists, like my friend Helen. Asked to recommend seniors who might qualify for a minority program in medicine, she went through her register and notified the appropriate students. When a female student came to her office the following day to inquire about the program, she suddenly realized to her horror that she had selected only males!

Teachers are members of our society, and they are just as likely as others to hold attitudes that are influenced by race and class. Some allow these attitudes to color their relations with their students. The educator Theodore Sizer sums it up: "Race and class snarl in many teachers' perceptions of students, leading to stereotypes. If you're black, you're poor. If you speak English haltingly, you're stupid. If you're white, you have a future. Blacks are basketball players. Blond is beautiful."[31]

High school senior Mina Choi set out to discover whether a student's race and sex influenced secondary school teachers' attitudes toward that student. Teachers in several schools were asked to rate twenty-eight black, white, and Asian high school students by examining their photographs. To conceal the true nature of the survey, they were told that it was a test of teachers' accuracy in predicting how students would achieve, and that the results would be correlated with each student's record. The teachers, most of whom were white, rated Asians by far the highest on math/science ability and motivational level, with blacks the lowest. The reverse rating applied to physical ability. Little difference was found in the comparative rating of females and males within each racial group.[32]

Other surveys bear out these findings. A regional study in the New Orleans public schools, completed in 1988, found that 56 percent of the teachers don't expect black males to go to college, and therefore do not encourage them to achieve in school.[33]

Julie (age 33), an African American, told of her experiences with math: "It could have been less difficult and stressful if teachers would have displayed a more optimistic attitude regarding expectations of students."

Let's look again at the Montgomery County, Maryland, report. For several years the county had offered sensitivity training to the staff and had placed special emphasis on participation of female and minority students in higher-level courses. Yet black students in honors classes felt that the teachers regarded them as inferior to white students. Some teachers accounted for the lower performance of blacks and Hispanics by blaming the home or society, problems about which they felt they could do nothing. On the other hand, black parents were most vocal in supporting the need for their children to study math.

In this report, teachers admitted that their expectations were different for students in the advanced track than for children working below grade level. The advanced groups were expected to think at a higher level, to deal with complex problems and new situations. Students in the low track were exposed to repetitious drill, rote learning, and the most simple examples.

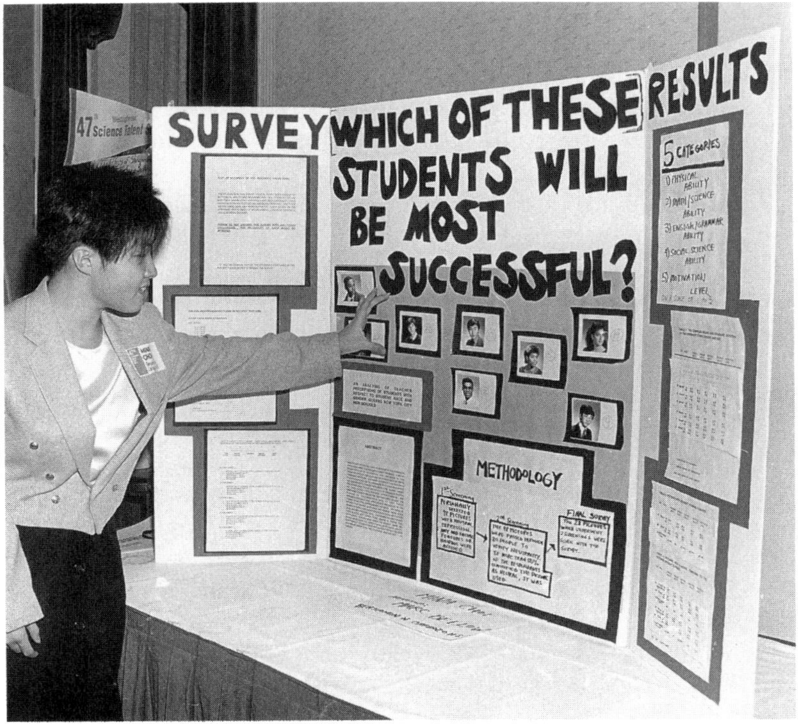

Mina Choi with her award-winning project in the 1988 Westinghouse Talent Search. (Photograph by the *New York Teacher.*)

Obviously the high- and low-level groups experienced math in different ways!

As for the attitudes toward female students, over half the principals and counselors believed that math was for men. They thought that female performance in math differed from that of males for the following reasons: lack of interest, the perception that math is not useful to women, and the belief that females are not as competent in math as are males. The old stereotypes persist. Furthermore, many elementary school teachers, most of whom are women, admit that they are not comfortable teaching math, and may, unwittingly, provide negative role models to their female students.[34]

Some of my respondents traced their fear of math to specific incidents involving elementary school teachers or classroom practices, experiences they still remembered years later:

- *Lisa (age 38):* "First grade, figuring a math problem, very easy, in front of class, forced by teacher to come to the chalkboard. I couldn't do it. I hated being in front of the class. The teacher became angry with me. I cried and stayed there in front of the class until she told me I could sit down."
- *Carla (age 25):* "In Grade Two I was always afraid that my peers would think I was stupid because when I went to the chalkboard I could never figure out what I should be doing about problems on the board."
- *William (age 24):* "When I was younger my teachers didn't care if I ever learned math at all. I am now involved in a learning-disabled study. I am now finding out what went wrong in retaining math."

Addie (age 38), a guidance counselor, is in a good position to analyze her own avoidance of math. In the small college she had attended, word went out that the math professor was mean and tough, and so she avoided taking any math beyond the introductory course. Now she regrets not having a deeper background in math for the educational measurements course she needs for an additional license. Looking back, she wishes she had encountered more interesting, effective, and competent teachers. "More positive attitudes about math would make the stigma [!] of taking math classes a lot less for my students that I counsel every day."

With the present shortage of mathematics teachers, schools may not be able to engage the "interesting, effective, and competent teachers" that Addie recommends, or the people of color who would serve as role models to inspire young Hispanics and blacks to continue their education. Teaching can be a hard, low-paying, unrewarding job, and no longer attracts the "best and the brightest." Young people, especially young women, now have more options. Schools, particularly in large cities, make do with the teachers that they can get. Often these people are, at best, untrained, and, at worst, uninterested.

Sharon and Anita are eighteen-year-old students entering California universities. Enrolled in summer bridge programs in math, they blamed their high school math teachers for their earlier failure. When Sharon's geometry teacher became ill, the class was taken over by about twelve different substitutes in turn. Some knew nothing about geometry, while others had their own peculiar ways of teaching. It was a wasted year. No wonder Sharon dropped math!

Anita wrote in her math autobiography: "The teachers didn't know themselves what they were teaching. We students corrected them. Instead of having a math class, I considered it a free period." And that is how she received credit for Geometry, Algebra II, and Mathematical Analysis.

Mark (age 20) had overcome his difficulties in learning algebra by attending the Pre-Professional Development Program at his California community college. The son of poorly educated Mexican parents, he received little concrete help at home and found the high school environment anything but inspiring. Some of his math teachers, licensed to teach physical education, were "learning from the students because they couldn't understand what they were doing. . . . I started to feel this hatred for education altogether. I felt like I was repeating elementary school." A low-paying job in Silicon Valley proved to be too monotonous, and Mark finally found his way into a college that met his needs.

Like these young people, you may also look back at your school experiences and wish that they had been different. "If the teacher had known the material better . . . if she had given me more time to answer the questions . . . if he had shown more interest . . . if, if, if . . . then I would not have developed this fear of math, this anxiety, the need to avoid anything resembling mathematics." But, as the math autobiographies attest, it is not too late to overcome these fears and anxieties. As a first step, you should identify those experiences and incidents that had a negative influence on your feelings about mathematics, as well as those that affected you positively. Above all, you should not blame yourself for the inadequacies that are due to conditions beyond your control.

Does Everyone Get an Equal Chance? Mathematics and Social Stratification

Critical sociologists of education, such as Michael Apple and Henry Giroux, claim that schools function to sort students according to the roles that they are expected to fulfill in later life. They argue that children of the advantaged social groups are educated to become the leaders of society, while working-class children are trained to engage in dull, repetitive work and to give blind obedience to authority.[35]

To investigate how this theory applies in practice, Jean Anyon of Rutgers University visited fifth-grade classes in northern New Jersey schools representing four different social strata: working class, middle class, affluent professional, and executive elite, to use her categories. In the working-class schools the student body was about 85 percent white, while the other schools were almost entirely white. She observed the children in all their subject areas and interviewed their teachers, all of whom had been rated "good" or "excellent" by their principals. I will summarize her general observations about the math curriculum, teacher attitudes, and classroom atmosphere.[36]

Working-Class Schools

Children were expected to follow the steps of a procedure, with little explanation of the reasons for that procedure or its relevance to anything else they had been taught. They copied the rules from the chalkboard into their notebooks and were evaluated on how well they followed these steps. The entire teaching process for division by a two-digit number consisted of the rules: "Divide, multiply, subtract, bring down." When some students seemed confused, the teacher said, "You're confusing yourselves," and just repeated the rules. The teachers made no attempt to explain the concepts or to use manipulative materials that might clarify the procedures.

The teachers would conduct oral drills in a manner that reminded Anyon of a sergeant drilling recruits, firing one question after another with no time for thought or discussion. They barked out orders without so much as a "please" or a "thank you." Students were not to leave their seats without permission. The students, for their part, did all they could to resist the teachers' orders and the required assignments. On the whole, very little actual teaching occurred. Perhaps some of these students were making a conscious effort to "not-learn," on the assumption that they had nothing to gain.[37]

Middle-Class Schools

Children were expected to get right answers, although some choice of procedure was permitted. When children reviewed homework, they were expected to describe their procedure. Lessons were based on textbooks. One child described school work in these words: "You store facts in your head like cold storage—until you need it later for a test or your job."

Professional and Executive Schools

Creative work was carried out independently. Children were expected to think about the concepts before arriving at answers to problems. They used calculators if they wished, and had access to a variety of materials. Children moved freely about the room and talked to one another. Teachers treated them with respect and held them responsible for their work. Students were trusted to check each other's work and to carry out experiments on their own, while teachers circulated to give help as needed.

What a contrast among these schools! It is obvious from these brief descriptions of teaching methods, classroom atmosphere, and relations of the students with their teachers that some students are being groomed to take power and others are expected to be subservient. Most of these students are white; imagine how racism aggravates the gap between the empowering and the authoritarian classrooms.[38]

Most of the math autobiographies I received indicate that the writers attended working-class schools, or, at best, middle-class schools. An elite school student who performed poorly in math would probably have received immediate attention and help, either from the teacher or from private tutors.

Before coming to the United States to enter college, Ron (age 20) had completed secondary school in Indonesia, where social class divisions are even greater than in this country. Not only did he attend private junior and senior high school, but "if we found some lessons difficult, we could take a private course. I took a private course in math, physics, and chemistry when I was in senior high school. . . . Sometimes I found something complicated and hard to understand. So I usually asked it of my private teacher. My dislike of math was overcome after I was taught clearly by my private teacher."

The eighth-grade mathematics curriculum provides another example of social stratification. In connection with the Second International Mathematics Study, eighth-grade students in eighteen countries were tested in 1982 on several aspects of mathematics. The United States did not show up well, ranking far behind Japan, France, and Hungary. Probing into the eighth-grade math curriculum, investigators found that schools in this country offer four dramatically different programs:

- Algebra, at the most advanced level
- Enriched—arithmetic, some algebra, a little geometry
- Typical—a great deal of arithmetic, a little algebra and geometry
- Remedial—predominantly arithmetic, review of elementary school topics

What criteria were used to place students in one or another of these tracks? When the investigators compared the arithmetic test scores of students entering the four levels, they found a large overlap among the levels, to the point where the lowest-scoring students in the algebra strand had lower scores than three-quarters of the students in the remedial classes. This sorting process has the most drastic negative impact on minority and low-income young people.[39] Remember Mark and his comment that high school made him feel like he was repeating elementary school?

Another aspect of social stratification is the placement of a disproportionate number of poor and minority children in special-education classes for math and other subjects. Presumably these children have been diagnosed as having learning disabilities (LD), biological conditions that prevent them from learning as normal children do. In practice, special-education classes are often filled with students who merely need remedial work. With federal funding available for special education but not for helping average students, schools take advantage of the situation. According to Gerald Coles, "special-education placement, especially for ethnic and racial minorities and lower-working-class children, is further demonstration that LD has functioned in the schools quite removed from professional decisions based on valid evidence."[40] And the stigma of the label "special ed" may stay with the child for a lifetime.

As for children considered learning disabled in math, Allardice and Ginsburg believe that "the most obvious environmental cause of poor mathematics achievement is schooling that is especially inadequate in the case of mathematics."[41]

They recommend the investigation of each child's learning *potential,* rather than reliance on achievement test scores. Attention should be given to the child's learning style and personality and to the social context of learning. Taking these factors into consideration "would be an important step in moving away from blaming the victim and toward acknowledging the complex interactions involved in learning."[42]

What about the person who has a genuine learning disability, a biological condition similar to dyslexia? There is much disagreement as to the causes and treatment, and it's beyond the scope of this book to make any specific recommendations. But there is hope. Just as dyslexics learn to read, so people learn to cope with learning disabilities in mathematics. The journal *Focus on Learning Problems in Mathematics* is devoted to the discussion of such problems and their solution.

The College Scene

The work force of the future will be better educated than it was in the past. To a far greater extent than at present, this well-educated work force of the future will include women, African Americans, Hispanics, Native Americans, and people from low-income and working-class families.

Colleges are now prepared to admit students who would not have dreamed of a college education a few years ago. Through no fault of their own, for reasons that we have already discussed, many of these students enter higher education inadequately prepared for college-level work. Colleges must furnish the remedial assistance that students should have received in their early years.

Most colleges now have "developmental" or "academic skills" departments to help students overcome their deficiencies. According to a 1988 estimate, 37 percent of college math courses are remedial. Even at Harvard, that elite, highly selective institution, some students need tutoring in decimal points and elementary algebra. But for colleges that serve students from the inner cities, the percentage of students requiring such remediation is much higher than the national average. Some institutions have summer bridge programs for entering students, to help them catch up before they become involved in the grind of full-time college work.

Many students need help in overcoming their fear of math, as well as remediation in mathematical content. Colleges now have programs to help math-fearers build confidence in their math ability. It might be just a one-session "math anxiety" workshop, a full course with a title such as "Mind Over Math," or a clinic staffed by a math instructor and a psychologist.

At the age of 45, Elsie was determined to overcome her fear of mathematics. She had lost confidence in her ability when a fourth-grade teacher told her she had no math skills. In spite of this hurtful remark, she received straight A's in high school algebra and geometry, for which she thanks a "very positive teacher who told us all we could do anything." With poor instructors in college statistics courses, "the old anxiety returned and it took me twice as long to pass the subjects as it should have." At the time she wrote her autobiography she was attending a math anxiety class and "attempting to overcome this anxiety."

For those who cannot or prefer not to attend full-time four-year colleges, there are workers' schools and community colleges. Early in the 1970s Wayne State University in Detroit set up the first college for worker education, a program to help working adults obtain their degrees by attending college evenings, weekends, and summers. By 1988 there were twelve such centers in the United States, and they had long waiting lists. Subjects are scheduled for large blocks of time; for example, students may concentrate on math one semester and science the next. Participants work in cooperative groups to help them overcome their fears and anxieties more easily. Working people need to be prepared as broadly as possible for a rapidly changing world, and workers' colleges provide one solution.[43]

Community, or two-year, colleges have seen tremendous growth in recent years, to the point where in 1989 they enrolled 43 percent of all students in higher education. Well over half were women, and the average age of all students was twenty-eight years. Because of the lower tuition costs and less rigorous entrance requirements than those of four-year colleges, two-year institutions are particularly attractive to working people with families and to low-income students. In 1986 community colleges enrolled almost six out of every ten Latinos/Latinas and American Indians attending college, more than four of every ten African Americans and Asian Americans, and over one-third of whites. Some eventually transfer to four-year institutions, while others have the goal of upgrading their work skills or pursuing new careers that require only an associate degree.[44]

Although women are attending institutions of higher learning in record numbers, they often find the college environment a chilly one. Women who have had clear sailing throughout elementary and secondary school may become discouraged by their college experiences. A report with the expressive title "The Classroom Climate: A Chilly One for Women?" details the many types of sex discrimination that may induce capable women to give up their career aspirations, and later research confirmed these findings.[45] Teacher behavior is often similar to that observed in the lower schools. Female students are often considered less capable and less serious than male students. Instructors involve men more than women in class interactions and permit men to interrupt women, while the female students themselves lack the self-confidence to stand their ground. Women may be asked purely factual questions, rather than the "higher-order" thought questions directed at men. Because a disproportionate number of college faculty members are men, women may not get the help they need. One analyst suggested: "College catalogues should carry warnings: The value you receive will depend on your sex."[46]

An instructor, trying to be helpful, may make a demeaning remark: "I know that women have trouble with spatial concepts, but I'll be happy to give you extra help." Not so innocent are the openly sexist jokes and the unsubtle remarks disparaging women. Whether subtle or overt, sexist behavior damages both women and men, as well as the whole educational process.

When Shiela Strauss attended Hunter College High School several decades ago, this public high school for the gifted was all female. Now that the school was coed and she taught mathematics there, it seemed to her that female students in the most advanced calculus classes were much more passive than in her day. To test her perceptions, she carried out a controlled experiment. Sure enough, the results bore out the fact that the girls were less self-confident and perceived themselves as ranking lower than the boys, although they performed as well as or better than the male students.[47]

In both high school and college, women attending single-sex schools take more courses in math and science and are far more likely to major in those fields than women in coed institutions.[48] At Spelman College, a historically black women's school, two out of five students were majoring in math, science, and engineering in the mid-1980s.[49] Indeed, when I attended all-female Hunter College, mathematics was second only to French as the most popular major field.

Historically black colleges and universities have been in existence for more than a century, established at the time of rigorously enforced segregation. A newer development is the growth of institutions and programs to serve Hispanic students. As of 1991, twenty-seven Indian tribal colleges, mainly two-year institutions, had been set up on reservations in Midwestern and Western states. Funding for most of these colleges and universities is far from adequate.[50]

The historically black colleges and universities are more successful than predominantly white colleges in producing African-American mathematics graduates. Although they granted one-third of the bachelor's degrees earned by black students in 1981, they were responsible for over half of the mathematics degrees and 40 percent of the degrees in computer and physical sciences earned by black students. It is interesting to note that a larger percentage of black women than men have majored in math.[51] Small, poorly equipped Xavier College in New Orleans is outstanding in fostering black premedical students; confidence building and cooperative work are the key ingredients in their success.[52]

Although the federal government has urged the desegregation of some of the state-supported historically black institutions, many African-American educators feel that they must be maintained. In a deposition concerning the future of the Louisiana state college system, Dr. Dolores Spikes, mathematician and president of Southern University, wrote: "Historically white colleges are not capable of addressing the needs of black students because whites are socially and culturally deprived of understanding the needs, desires, abilities and mores of black students."[53] The hostile climate and outright racist incidents at many predominantly white colleges have discouraged blacks from attending these institutions.[54]

Both the enrollment and retention of minority students is a serious issue that is finally receiving some attention. Poverty and inadequate financial aid are problems that affect all low-income and many middle-income people, and minorities are especially hard hit. Outright grants have become scarce, and

Teachers as Motivators: Clarence Stephens and Lee Lorch

It is deplorable that few women and people of color are on the faculties of mathematics departments outside of single-sex and race/ethnic-related colleges and universities. Yet we can point to programs and instructors that were successful in developing mathematicians among women and people of color in spite of the absence of role models.

One such success is the department of mathematics at the State University of New York College at Potsdam, formerly headed by Dr. Clarence Stephens. Only one of the sixteen faculty members is a woman. Yet, as John Poland writes in his glowing description of the program, "they graduate more women in mathematics than men. . . . The female mathematics students attribute their success to the supportive atmosphere."

Another outstanding example is Dr. Lee Lorch, a white professor at historically black Fisk University. During the five years (1950–1955) he headed the mathematics department, he "influenced one-fourth of [his students] to pursue and earn the master's degree in pure mathematics. Moreover, one-tenth of the students continued to the doctorate. Each known doctoral recipient credits Lee Lorch as the greatest influence in his choice of career," wrote Dr. Vivienne Malone Mayes, a former student. His warm concern for the welfare of African-American people prompted an attack by the Congressional Committee on Un-American Activities and dismissal from Fisk by the white Board of Trustees, without charges or trial. To this day he has continued to work on behalf of minorities and women. (John Poland, "A Modern Fairy Tale," *American Mathematical Monthly* 94 [1987]: 291–295; Vivienne M. Mayes, "Lee Lorch at Fisk: A Tribute," *American Mathematical Monthly* 83 [1976]: 708–711; "Citation for Lee Lorch," *Newsletter* [Association for Women in Mathematics] 22, 2 [March–April 1992]: 5–6.)

SOURCE: CAROL FRANCES & ASSOCIATES/U.S. DEPARTMENT OF EDUCATION.

students are reluctant to take on a heavy debt load. These young people might well ask why tuition soars and grants disappear just when more people of color hope to go to college. Our national leaders say that well-educated people will be needed in the work force of the future. How are they to get their education?[55]

Another problem is the decline in the number of students majoring in mathematics, just when more fields are becoming mathematicized. Who will enter these fields? Who will teach the coming generation? Who will be the role models in the future? The problem is even more serious for minority students, who suffer from inadequate high school preparation, poor counseling, a shortage of minority math and science teachers, and too few role models and mentors to spur them on.

Reform at all levels is essential. In the words of a recent report: 'Most of the interventions devised by colleges and universities are aimed at enabling students and/or faculty from underrepresented groups to fit into, adjust to, or negotiate the existing system. . . . [Only by] changing the climate in which the students are educated can we . . . significantly affect the participation of women, minorities, people with disabilities, and, indeed, all students in science, mathematics, and engineering."[56]

Yes, change the climate, reform precollege education, grant more scholarships, and, most important, encourage young people to complete their education by offering the tangible reward of a good job at the end of the tunnel.

CHAPTER SIX

School Math Is Not Necessarily Real Math

I did not dislike math in elementary school and did well, as I recall. There was something satisfying about doing arithmetic problems, but with hindsight, I realize I never really went beyond *performing* and not developing any real number sense—how these different types of computations relate to one another.

With hindsight, I can understand that my lack of confidence and eventual fear of things *mathematical* were due in great part to lack of experience and never having had an opportunity to develop shortcuts for calculations. I tended to try to do mentally what I'd been taught to do on paper and with pencil—borrow from the ten, carry the one, etc.—never rounding off or estimating generally.

Can you imagine what it was like to study statistics as both an undergraduate and graduate student? I survived those courses—without understanding—by memorizing and plugging in formulas at the appropriate times. In response to my questions about how to decide *what* statistic to apply to my research data (not just how to apply the statistic as given by the professor), I was told I needed calculus for that. Oh, great! I didn't get the algebra in all I was doing; how could I grasp calculus?

—*Mary Jo Cittadino, Mathematics Educator, EQUALS, University of California at Berkeley*

Hassler Whitney was one of the most prominent and creative mathematicians of the twentieth century. A long-time professor at the Institute for Advanced Study in Princeton, New Jersey, he spent his later years in elementary classrooms, observing how math is taught and teaching classes himself. He deplored what was happening in the field, fearing that math avoidance would increase as a result. When adults told him, "I never could do math," he sympathized with them and responded: "You never had a chance to see or do real math, which is easy and fun."[1]

Elementary Math Can Be Real Math

I will start this section on an upbeat note by telling you about the "real math" I witnessed on a visit to my friend Judy's classroom.[2] She teaches a combined third- and fourth-grade class of socioeconomically and culturally diverse students in an urban school. The children had just finished a unit on Africa. Patterned cloth from several regions of the continent decorated the room. Neatly labeled African objects lay on an open shelf. On a large table was a miniature circular compound of round houses, some fashioned from tiny clay bricks, others molded from clay, their conical roofs covered with foil to simulate metal. A shelf held a collection of *Owari* game boards the students had made from styrofoam egg cartons and painted in brilliant colors. *Owari* is one version of the African mathematical game known by the generic name "mankala," and rated one of the ten best games in the world. Two players take turns transferring seeds or pebbles around the board in order to capture the majority of the playing pieces.

Judy invited me to examine the children's portfolios, stacked in a tall pile on top of a bookcase. It was obvious that mathematics was an important component of their work. They had drawn geometric "game boards" for several African versions of three-in-a-row games. Before constructing their compound of round houses, they had carried out an experiment to determine the shape that would have the largest floor space for a given amount of materials for the walls. Using a fixed perimeter of sixteen units, measured by a sixteen-unit length of string or strip of grid paper, each child had drawn a circle and several rectangles with different dimensions on a sheet of grid paper. Counting the grid units enclosed by each shape convinced the children that the circle had the greatest area, while the area of the rectangle diminished as the two dimensions grew further apart. In the course of this hands-on experiment the students learned to distinguish between perimeter and area, developed the formulas for the perimeter and area of a rectangle, and understood the use of linear units to measure length and square units to measure area.

Why do some people live in round houses, while others prefer a rectangular shape? Why do some people construct flimsy temporary dwellings while others build solid structures to last a long time? The children had engaged in a spirited discussion of

Owari game board on *kente* cloth, both from Ghana. (Photograph by D. W. Crowe.)

these questions. They considered such factors as the climate, the available materials, the way the people made their living, and the traditions of the society. This discussion led to an investigation of the kind of homes and buildings they saw in their own communities. How would rectangular furniture fit into a round house?

These kids were solving real problems as they learned about other cultures and made comparisons with their own environment. Their study of African societies involved math, social studies, art and architecture, language arts, music, and science—just about all their school subjects. Various branches of math came into play—arithmetic, geometry, measurement, decision making.

Students of varied "abilities" worked in cooperative groups and discussed difficulties as they arose, while the teacher or an assistant circulated, throwing out hints to stimulate children's thinking. All responses are accepted. There is no punishment for wrong answers, but children are expected to justify their conclusions. Writing is important; students wrote their questions and procedures in their notebooks. The project offered a variety of activities to engage students of diverse backgrounds and learning styles. General class discussion often followed the group activities.

Discipline is hardly a problem; the kids are too involved to get into trouble! While Judy was showing me the portfolios and exhibits, the children were working in their groups to solve the "problem of the day," an algebra problem about the purchase of various quantities of East African cloth and shoes. The groups were using different strategies—guess and check, make a table, write algebra equations.

Judy manages to have plenty of help. The day I visited she had an intern from a local college (he just happened to have spent seven years in Africa), a parent, and a senior citizen volunteer who kept track of the clerical work, like checking that the students had completed their assignments. Another nice feature of Judy's class is that the children learn to count and to speak some expressions in Spanish and in Haitian Creole; the latter language is emphasized in the school.

Teachers like Judy put their students in control of their own learning. They expect that all children are capable of doing math, and trust them to work out solutions to real problems, both on their own and in cooperation with their classmates.[3] School learning is not separate from the outside world; school, community, and the world beyond are integrated into a meaningful whole.

Your experience with school math was probably radically different from the picture I painted of Judy's classroom. Even today, the majority of students are not enjoying the "real math" that would be meaningful to them. In the early grades children are expected to learn mathematical words and symbols, and to master paper-and-pencil procedures involving operations with numbers. Since they see no connection between these procedures and the real world, it all makes little sense to them. The only way they can learn the required material is to memorize it, bit by bit, as it is fed to them.

With the emphasis on accountability, measured by standardized tests, teachers feel compelled to use methods that they know are harmful. My friend Paula asked me how she can get her first graders to memorize the addition facts. In the past her students had learned by experimenting with various kinds of manipulative materials—blocks, beads, Cuisenaire rods, to name a few. With the new mandate from above—rigid curriculum

Mind Your Language!

Suppose Cathy has five tapes and Kwame has twenty tapes. I might compare the numbers by making any of these statements:

- "Kwame has fifteen more tapes than Cathy." I look at the difference between the numbers.
- "Kwame has *four times as many* tapes as Cathy." I multiply Cathy's number by four to get Kwame's number.
- "Kwame has *three times more* tapes than Cathy." I look at the difference between the numbers—fifteen—and see that it is three times Cathy's number.

But the last statement is probably wrong, although it might have been correct a few years ago. The style seems to have changed. Here is an example from the *Los Angeles Times* (11 March 1991, "In Brief," page B3). Regarding asthma, Dr. Sheffer is quoted as saying: "The mortality rate for Afro-American males is four times that of whites, and for [black] females it is three times higher."

I judge from the context that "four times the rate" is greater than "three times higher than the rate." I would have thought that both expressions have the same meaning. Obviously they don't. I am confused!

and precise time frame to "prepare for the test"—there was no time for "frills" such as concrete experiences with numbers. The main goal was to raise the test scores in the district, which ranked near the bottom in the city. I convinced her that she must resist the pressures if she didn't want the children to grow up thinking that math is just a matter of rote memorization and believing that they were incapable of taking charge of their own mathematical learning. Subsequently a delegation of teachers from the schools in the district was able to persuade the administration to relax the pressures somewhat.

Schools generally ignore the fact that children enter school with a considerable amount of mathematical knowledge, picked up from their families and the environment, and processed by means of their own problem-solving abilities, just as they acquired language. Children learn to talk because they need to ask for something, to do something, to relate an important event. They don't need to be "taught" speech. They learn by watching and listening to other people. Praise and encouragement from adults help children to develop self-confidence and the motivation to take on further challenges. Young children love to learn new ideas. School so often squelches their desire to learn.

Just imagine this situation. A mother tells her two-year old, "Now I am going to teach you to talk." She proceeds to spend a half hour every day for a month having her child memorize nouns: "Mommy, Daddy, truck, cookie," and so on, giving her or him the words that *she* thinks the child should know. During that period the child is to practice nothing but the given nouns. The mother ignores anything the child might say the remainder of the day and any words that are not on the list. The following month she "teaches" the child verbs. Now she drills her or him on specific verbs and nouns during the daily half-hour lesson on "learning to talk," but continues to pay no attention to the child's speech at any other time or to words that she hasn't "taught."

Ridiculous, isn't it? But if you substitute "math" for "talk," "teacher" for "mother," and "addition and subtraction" for "nouns and verbs," you have a picture of the "teaching" that goes on in many early-grade classrooms. But teachers are not at fault. It is our whole factorylike educational system that is at fault.

I still recall my feelings of complete confusion in second grade when I returned to school after a three-week illness and found the class doing subtraction. I had figured out my own methods of working with numbers—numbers held no fears for me. What I found bewildering was the terminology the teacher was using and the procedures that we were to follow in our written work. I soon caught on, but I can imagine how frustrating it must be for a student to find himself or herself in a permanent state of bewilderment about school math.

An experimental (at present) early childhood mathematics program at the University of Wisconsin rests on the premise that kids enter school knowing quite a lot of math! With this approach, teachers in the Cognitively Guided Instruction (CGI) program listen to the children, rather than trying to impart a prearranged body of knowledge. They make decisions about the kind of problems that each child can handle on the basis of the child's responses. Children use a variety of strategies and are expected to justify their thinking. Math happens at any time of the day and in connection with a variety of activities. As for test scores, these children far outrank those students who spend a good deal of time on routine drill and practice.[4]

Many of my respondents recall their early experiences as rote memorization of the multiplication tables and other arithmetic facts. Connie (age 22), finishing college with honors, recalls a fourth-grade teacher who hit the children with a ruler when they made mistakes in their multiplication facts. To this day she has no confidence in her ability, in spite of her excellent high school math grades.

Years after completing school, Margaret still felt guilty about her inadequacies in math. She remarked to her college algebra teacher: "Whenever I come to your class I feel like I am going to vomit. I never learned my multiplication tables and I feel so guilty about it that it makes me sick." With some reassurance from the instructor and a little help from her husband, she quickly filled in the gaps.[5]

Not only was rote memorization of facts the main goal, but these facts had to be regurgitated fast! Flash cards and the need for speed were responsible for many a stomachache. But John (age 24) loved them. In the elementary grades, "math was fun because of its ease. Flash cards made learning more fun." A good example of the male competitive spirit!

Another bugaboo of elementary math is long division. A mother and her sixth-grade daughter were waiting with me in the doctor's office. For a half hour the girl worked in a desultory fashion on her math homework, a page of division exercises, then pleaded with her mother to be allowed to finish at home. Her mother was adamant that she must continue with the pencil-and-paper computations. As I was leaving I looked at the instructions in her textbook: "Estimate the answers, then check with your calculator." Why wasn't she following the instructions? "My mother won't let me use a calculator," she complained. Not only was she failing to learn the important skill of estimation, so useful in real life, but she was learning to hate math.

An incident in my college class for future teachers brought

home to me the power of the teacher as authority, not to be questioned. I had posed this situation: "Suppose that your students have learned how to multiply a two-digit number by a one-digit number. Now you would like them to multiply by a two-digit number. Form groups and come up with a lesson plan."

I asked for a volunteer to teach the lesson, while the rest of us pretended to be fourth graders. Beth was the brave one, and proceeded to demonstrate with the product of 34 and 12:

$$\begin{array}{r} 34 \\ \times\ 12 \\ \hline \end{array}$$

"Two times 4 is 8, and 2 times 3 is 6." She wrote:

$$\begin{array}{r} 34 \\ \times\ 12 \\ \hline 68 \end{array}$$

"Then you multiply one times 4, and write the 4 under the 6," and so on—an absolutely correct procedure.

$$
\begin{array}{r}
34 \\
\times\ \ 12 \\
\hline
68 \\
34 \\
\hline
408
\end{array}
$$

All the students said that they understood perfectly, with one exception—me. I raised my hand and asked: "Teacher, why did you put the 4 under the 6, and not under the 8?"

Immediately every young woman (this was a single-sex school) turned on me with the cry: "But that's how we were taught!" They assumed that I was challenging the correctness of the procedure, rather than the concepts that underlie the procedure. It had never occurred to them to ask "why?" Their teachers were the authority, the last word, not to be questioned.

Many students learn "how" but not "why." Jeanette (age 46) wrote about her college math courses: "I had long since resigned myself not to understand very much, and as long as I did this and just memorized the appropriate information, I did very well—even got A's and B's!" It was only after she had become an elementary school teacher that she began to understand the concepts that had eluded her in the past. In order to "explain the whys of math, I had to read different explanations of what it all meant, and how to proceed step by step." As a result of her self-teaching, math doesn't scare her any longer, and she can deal confidently with her students.

Ruth (age 42) is also an elementary school teacher. Her story is just the opposite of Jeanette's. "Failed math in fifth grade—total lack of skills—plus fear—inability to to comprehend *any* math—rote learned it." She feels okay teaching up to sixth-grade math, but "total freeze if I'm being taught something new." She can't think of any way to overcome her fear. "Math makes *no* sense—especially when I go absolutely blank if someone tries to teach me. Even discussing math makes me forget everything."

With constant drill, children usually become adept at computation. Solving real problems is another matter. Textbook "story problems" are worded in such a way that children learn to look for key words. For example, the word "left" in a problem

Estimate and Live!

The ability to estimate might have saved lives. According to an item in the *New York Times* (29 August 1987, p. 32):

> A tractor-trailer driver looking at a calculator instead of traffic triggered the fiery chain reaction collision on the New Jersey Turnpike near Cherry Hill that killed six people and injured thirteen, the authorities said today.
>
> The truck driver . . . was driving between sixty and sixty-five miles an hour, calculating his mileage, when his truck rammed into a car and plowed through seven other vehicles.

The driver had been hauling peanut butter in a rented truck. He was in stable condition with leg and face burns.

Suppose you wanted to estimate your mileage per gallon of gas. What is the procedure? Let's say you drove 487 miles and used 19.3 gallons of gas. You need to divide 487 by 19.3. You can get a good enough estimate by rounding the numbers to 500 miles and 20 gallons.

$500 \div 20 = 50 \div 2 = 25$. Your gas mileage is about 25 miles per gallon. Now try some examples of your own.

means subtraction. It isn't even necessary to read the whole problem; just subtract the smaller number from the larger number. Besides, every problem on the page is done exactly the same way. No need to think at all! It doesn't take long for children to internalize the idea that understanding has no relationship to school math.[6]

Children look for certain clues in the wording of the problem or examine the type of numbers given, to guide them as to whether to add, subtract, multiply, or divide. But even when they do use the correct operation, the answer may not make sense. A problem similar to the following appeared on a national

test: "A school is ordering buses for a trip. Each bus can hold fifty people. How many buses will be needed for 125 children?" Many students gave the answer 2½, or 2 with the remainder 25. The context of the problem was completely irrelevant. Did they visualize 2½ buses?

Emma (age 45) is an organizer in the educational community and a self-taught artist. For her, the context is most important. As a child in Puerto Rico, she could not accept math as presented—rote memorization of facts and procedures. But her questions went unanswered, and she finally resigned herself to giving the expected responses. She ran into real trouble in a New York City junior high school. Math tests had to be written neatly in ink, with no crossing out or erasures. "The teacher picked on me a lot because of my poor English." In high school she was assigned to "Record Keeping" instead of mathematics. Thus ended Emma's encounter with school math.

After working and going to school part time, Emma received a scholarship to a prestigious liberal arts college to study psychology. Because she was a top student, the one-course math requirement was waived. Eventually she earned a doctorate in Human Relations, specializing in race, class, and gender issues in education, a program with no statistics requirement. Although she has no trouble with budgets and shopping, she avoids analyzing graphs, for example, when she studies a report, preferring to read the figures. "I imagine my own graph. . . . I guess if I had been interested in math, I would have found some creative ways of dealing with the subject."

Textbook word problems give just enough information to enable the student to arrive at the one and only correct answer, using the one and only correct method. Real life is seldom so accommodating. When we handle real problems, it may be necessary to eliminate extraneous data or to do further research. How do we deal with facts that contradict one another? Mathe-

Same Numbers, Different Answers

The following examples all involve the same numbers and the same operation. But the answers are all different.

1. Divide 125 by 50.
2. 125 employees are going to a picnic by bus. Each bus holds 50 people. How many buses must be ordered?
3. A carton of milk costs 50 cents. How many cartons can I buy if I have $1.25?
4. Fifty pieces of ribbon, of equal length, are to be cut from a spool of ribbon containing 125 m (meters) of ribbon. Find the length of each piece, to the nearest centimeter (cm). (One meter equals 100 centimeters.)

Answers:
(1) 2, remainder 25; or 2½; or 2.5
(2) Three buses
(3) Two cartons of milk
(4) 2 m and 50 cm; or 250 cm

In response to a question similar to the one about the number of buses, a student in a low-income neighborhood suggested that the extra passengers squeeze onto the smaller number of buses to save money. ("K–12 Testers Offer Poor Defense of Multiple-Choice Items," *FairTest Examiner* 6, 1 [Winter 1991–1992]: 7.)

maticians often have to experiment with several different methods of solving the problem before them. Real math is never so cut-and-dried as the textbook implies, or as presented by many teachers, especially college professors.

School math may even get in the way of solving real problems. At a summer workshop for elementary school children, I set out an open square box of pasta shells. The participants were to guess how many shells were in the box, and then place signed slips of paper with their estimates in a closed container. At the end of the afternoon several children counted the shells. Most

Word Problems, Word Problems

This problem, by Gary Hendren, was labeled "Special" and published on the Problem Page of the Missouri Council of Teachers of Mathematics *Bulletin* (March 1992). Don't be alarmed. The introduction to the problem stated: "This problem is meant to be read aloud for fun rather than solved."

> A rope over the top of a fence has the same length on each side, and weighs one-third of a pound per foot. On one end hangs a monkey holding a banana, and on the other end a weight equal to the weight of the monkey. The banana weighs two ounces per inch. The length of the rope in feet is the same as the age of the monkey, and the weight of the monkey in ounces is as much as the age of the monkey's mother. The combined ages of the monkey and its mother are thirty years. One-half the weight of the monkey, plus the weight of the banana is one-fourth the sum of the weights of the rope and the weight. The monkey's mother is one-half as old as the monkey will be when it is three times as old as its mother was when she was one-half as old as the monkey will be when it is as old as its mother will be when she is four times as old as the monkey was when it was twice as old as its mother was when she was one-third as old as the monkey was when it was as old as its mother was when she was three times as old as the monkey was when it was one-fourth as old as it is now. How long is the banana?

Now try an easier problem!

off the mark in her estimate was the invited guest, a college administrator. She had used—or misused—a dimly remembered formula and arrived at a fantastic result. The kids with the best answers explained their procedure: Count the shells in the top layer, estimate the number of layers, and multiply.

Temperature Conversion Made Easy

The United States is the only major country that has not yet converted to the metric system in practice. When you're in Canada and you hear that the temperature is ten degrees, or an announcer in the Caribbean tells you it's in the low thirties, what does it mean? You can obtain a good enough approximation by carrying out this estimation method mentally.

Estimation Method:

- If Celsius temperature is given, double the number and add 30 to arrive at Fahrenheit temperature, the scale used in the United States.
- If Fahrenheit temperature is given, subtract 30 and divide the answer by 2 to arrive at Celsius temperature, the scale used in most countries.

Let's test the method by starting with 10° Celsius:

$10 \times 2 = 20$; $20 + 30 = 50$. 10° Celsius = (approximately) 50° Fahrenheit.

How does 30° Celsius translate?

$30 \times 2 = 60$; $60 + 30 = 90$. 30° Celsius = (approximately) 90° Fahrenheit.

As a check, change the Fahrenheit temperatures back to Celsius.

Suppose you forget which operation comes first. That happens to the best of us. Remember that on the Celsius scale water freezes at 0°, and the freezing point on the Fahrenheit scale is 32°. Try out the procedure you think is correct by converting the freezing temperature from one scale to the other. If it doesn't give you the right answer, make adjustments in the procedure until it works—good practice in mathematical thinking.

Here are the accurate formulas.

Let F stand for Fahrenheit temperature. Let C represent Celsius temperature.

$F = (1.8 \times C) + 32.$ $\qquad C = (F - 32) \div 1.8$

(*Continued*)

Temperature Conversion Made Easy
(Continued)

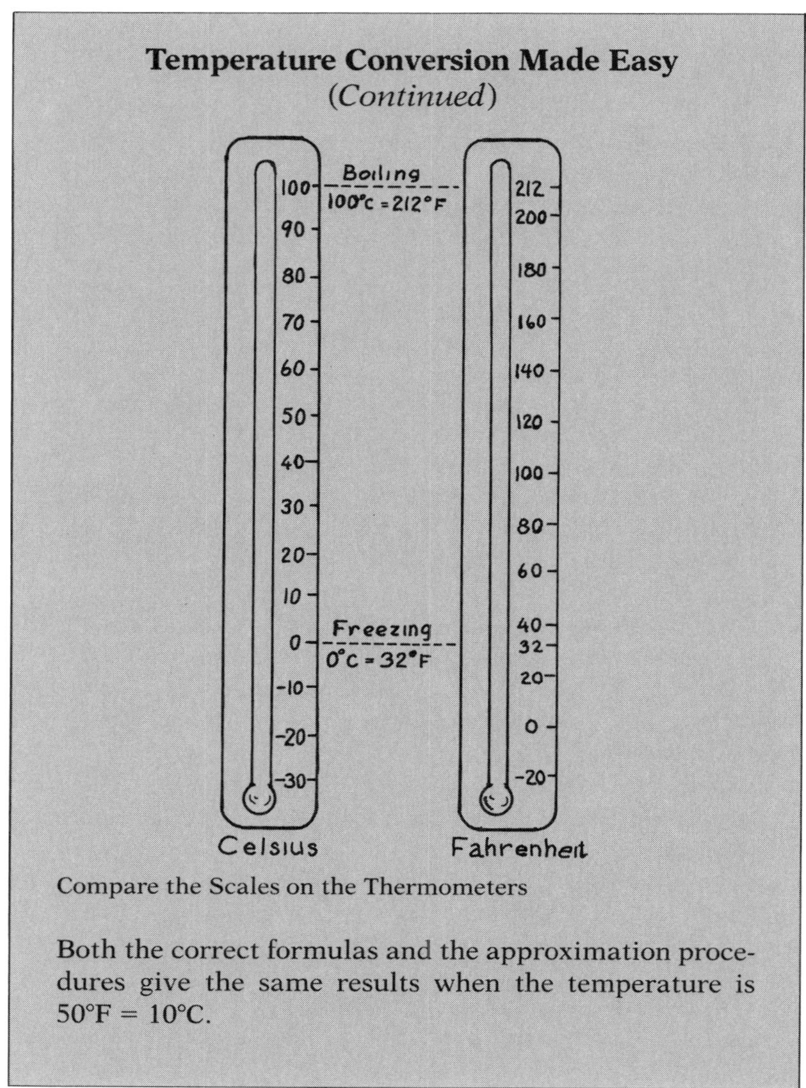

Compare the Scales on the Thermometers

Both the correct formulas and the approximation procedures give the same results when the temperature is 50°F = 10°C.

The Nemesis: Algebra, Geometry, and Calculus

Henry (age 17), a student in a summer bridge program, met his nemesis in seventh-grade algebra. His failure in the course was the beginning of his dislike of math. "It was too many rules for too many different problems. In high school I was put in 'shop math,' meaning that's the lowest in school." He finally took the first-year algebra course in eleventh grade. "I never was good in math, and I hate it with a passion."

Bob Moses's Algebra Project is designed for students like Henry. Back in the sixties, Bob Moses left his job as a math teacher to become one of the foremost activists in the struggle for the right of blacks to vote in Mississippi, and to organize the Freedom Schools in the South. Moses became involved again when his daughter found that algebra was not available at her middle school. He began to work with the students, teachers, and parents at the school to introduce algebra to all students instead of just a select few. The Algebra Project developed from these beginnings and is now becoming part of the curriculum in several large cities. Starting in sixth grade, the curriculum builds on life experiences to make the subject accessible to young people who might not be considered likely candidates for abstract math. Imagine learning about directed (positive and negative) numbers by riding the transit system to the city center and back![7]

The traditional high school geometry course was supposed to train people in deductive logic. It failed dismally with many students. "Better to teach logic with minute mysteries," remarked a perceptive high school student. Vivian (age 17) considers geometry "the worst man-made subject ever. The proofs were murderous, as well as the word problems. Everyone that has taken geometry says that they hated the proofs and the word problems." Vivian is right about geometry being "man-made." The ancient Greek philosophers started it. With their reliance on slave labor to give them leisure time, they engaged in the type of argument that gave rise to the proofs that bedevil high school students today.

Not only do many students learn to hate geometry proofs, but they even lose their ability to solve geometric problems by using their informal everyday knowledge, according to Dr. John Volmink of Cornell University. Children aged 10 to 16 were given several geometry problems and were permitted to use concrete materials to solve them. The younger children displayed greater creativity and flexibility than the older ones, who had already taken the standard high school geometry course. Volmink concluded that the spontaneity and self-confidence shown by

younger children may be suppressed after they have studied formal school geometry because these powers are never called into action.[8] With the technology that is now available, the nature of geometric investigation is changing. Using up-to-date calculators and computers, students can take the initiative in testing various hypotheses about geometric relationships.

An experimental secondary mathematics program is under development in California. Called The Interactive Mathematics Project, it encourages students to investigate a variety of real-life or imaginary situations from the mathematical point of view and at the same time engages them in many branches of mathematics—algebra, geometry, number theory, probability, logical decision-making. The titles of the projects are suggestive of their content: The Game of Pig (a probability game), The Pit and the Pendulum (based on the Edgar Allan Poe tale), Do Bees Build It Best? (what is the best design for a honeycomb?), and Leave Room for Me! (a study of the growth in the world population). For more information, contact EQUALS (see "Resources" at back of book).

Enrolling in a college math course after a long lapse can be an overwhelming experience. Beverly (age 27) had dropped out of high school after tenth grade. "I was introduced to algebra in college. It is like learning a new language for me. It just goes too fast and I don't get enough repetition to really learn it. I like to understand thoroughly before going to the next step. I find college too fast-paced for that. Seems like I don't really grasp one concept before we are on to another."

Jane (age 20) loved math in elementary and high school. She was even considering a math major—until she took college algebra. "My dislike for math began after I entered college because they no longer took their time teaching the subject step by step. Instead it was a short cut here and a short cut there. I never overcame my dislike for math; instead I became content knowing that I passed it and would never have to be bothered with it again."

By the time students take calculus they are expected to be able to read the textbook. In my day we called our text a "cookbook." Unfortunately for most students, "cookbook calculus" is still being taught. Anyone who can read a recipe should be able to read the calculus text and follow the directions. They need only memorize the symbols and definitions! The why and the wherefore of this very dynamic and useful subject remain a mystery to many students. Although I had been a straight-A math major, I never really understood the calculus until I studied it again as a reentry student twenty years after college graduation.

Martha (age 24), a premed college graduate, remembers her enjoyment of precollege math in Haiti. "The teachers used methods that made the problems fun, and we did not have to memorize the formulas." Her experience in a New York college was just the opposite. "The difficulty I had with calculus was mostly due to the memorization of a lot of formulas which were involved in solving the problems."

Marni (age 24), an elementary school teacher, had been a great math student until she got to calculus in college.

> In college I started Calculus I. I could understand all of the concepts as it was mostly a review of high school [calculus]. However, this is a known "weeder" course. I felt, even though I knew the material, that there was no way I could compete with the pre-med students and engineers. I became very stressed. After the first exam I dropped the course, never to take another math class.
>
> I think that a real help would have been in college, if there had been less emphasis on "weeding" people out of math and more on keeping people (expecially women) in. My T.A. [teaching assistant] did nothing to keep me from dropping (not even a word). I think if places like

Words into Algebra

A kind of test problem has been posed to many groups in several countries. It sounds simple, yet a large portion of high school and college students, as well as college faculty in nonmathematical fields, gave wrong answers. Problem: North Star College has six times as many students as professors.

- Let S represent the number of students.
- Let P represent the number of professors.
- Write an algebra sentence to show the relationship between students and professors.
- Which sentence is correct: $6S = P$ ($6S$ means $6 \times S$) or $6P = S$???

One way to check is to replace the variables S and P with some reasonable numbers. Go back to the first sentence in the problem: "six times as many students as professors." That means one professor has six students. Start with the sentence: $6S = P$. Substitute 6 for S, and one for P.

The sentence becomes: $6 \times 6 = 1$. This is obviously a contradiction! Work with the sentence: $6P = S$. Replace the P with one, and the S with 6.

The sentence becomes $6 \times 1 = 6$. This is correct.
Therefore $6P = S$ is the correct algebraic sentence.

Now for a real situation involving the student-faculty ratio. According to a story in the *New York Times* (19 October 1988, B8), Spelman College, a historically black women's college of 1,750 students, was considering plans to decrease its enrollment by 100 students. School officials wanted to "keep its faculty-student ratio at an average of 15 to one."

This reporter got it wrong. He obviously meant 15 students for each faculty member. He should have said: "keep its *student-faculty* ratio at an average of 15 to one." The order of the words "student" and "faculty" tells the order of the ratio "15 to one."

> [my university] had all women's sections with women T.A.s, they could make women students feel more comfortable and perhaps more would pursue math and science courses and careers.

Some college mathematics departments are aware of the issue and are trying to do something about it. One solution is to humanize the way calculus and other college math courses are taught. The mathematics department of the State University of New York College at Potsdam, under the chairmanship of Dr. Clarence Stephens, won accolades for doing just that. Instructors imbue the students with the confidence that they can learn the subject, provided that they are willing to work at it. During most of the class period students work together in informal groups, discussing and arguing about the concepts and procedures. The instructor is regarded as a coach, not as the authority, and is available for outside help when needed. During the period 1985 to 1987, an average of 24 percent of the graduating classes were mathematics majors, in contrast to 2 percent for the nation. Stephens, an African American, had taught formerly at Morgan State University, a historically black institution in Maryland, where he inspired many African-American students to continue in the field of mathematics.[9]

When Kunzen (age 18) wrote her math autobiography, she was studying advanced calculus while working as a tutor and draftswoman. In spite of her claim that she "didn't have an aptitude for math" in elementary school, she learned to love the subject in high school when she "figured out the trick to being good at it. You have to learn the concepts." How right she is!

Learning Mathematics in the Contemporary World

People tend to think of mathematics as a frozen body of knowledge, put together sometime in the past by a few geniuses, mainly white males of European heritage who had "mathematical minds." This impression is way off the mark. Mathematics is an ever-evolving field, and even young people can contribute to it. A good example is the annual Westinghouse Talent Search among high school seniors. Every year students come up with original research projects in mathematics and science.

SCHOOL MATH IS NOT NECESSARILY REAL MATH

One need not be a genius to do original mathematics. Children—and adults, too—often use their own invented strategies to solve problems, regardless of the methods they are supposed to have learned in school. Young children rely to a great extent on counting, including finger counting, a practice that is no longer taboo. An invented strategy is a person's own process, is based on that person's understanding. Naturally, it is more meaningful than a memorized procedure, something that is soon forgotten because it is barely understood. People who devise their own methods are doing real math!

How can math be presented to all people so that it's both meaningful and interesting? Why are out-of-school programs successful in involving the very same students who find school math a boring experience?

Listen to this Latina student talk about the contrast between her four-week summer program on math, science, and sports for "average" minority junior-high girls, and her school experience: "In Eureka science we get to do experiments every day and discuss and help our peers, but in school science you can't talk among your friends about the work or you will get into trouble. . . . You can't experiment every day in school because you are supposed to cover a certain amount of work by the end of the year."[10]

In a ground-breaking 1989 report, the National Council of Teachers of Mathematics (NCTM) voiced tremendous concern about the implications of the poor math performance demonstrated by students in the United States, and called for a complete transformation of school mathematics at all grade levels. Furthermore, the processes of evaluation must assume entirely new forms to free mathematics education from the "tyranny of testing." Other reports also dealt with the need for reform and made specific recommendations for the revision of both school and college mathematics education.[11] Their proposals include these goals:

- Students should be able to think mathematically to solve real problems. A problem involves a situation for which no solution is immediately apparent. Students should be able to make conjectures, see relationships among different aspects, apply their skills to carry out computations, check their results, and explain the reasoning behind the procedures. In other words, they must develop their higher-level thinking skills.

- A variety of experiences should be incorporated into the curriculum to enable students to develop those higher-level thinking skills. People learn by constructing their own knowledge. They extract new information from their experiences and relate it to what they already know, building to a higher level of understanding.
- At all levels, mathematics education must embrace such topics as algebraic thinking, computation, measurement, geometry, statistics, probability, mental arithmetic, estimation, and approximation.
- Students should become familiar with applications of mathematics, and be able to see the relationship of math to other subject areas.
- All students should have experience with calculators, computers, and a variety of materials for exploring mathematical concepts. In a move to have students spend more time learning math concepts and solving challenging problems, Chicago public schools began in 1988 to provide free pocket calculators to all students in grades four through eight.[12]
- The learning climate should incorporate high expectations for all students. They must feel free to ask questions and take risks, without fear of punishment for wrong answers. They should not be held back on the grounds that they have not mastered the so-called "basic" skills of computation. They should have opportunities to work with other students and to discuss their work.
- Teachers must be educated to handle the new curriculum and methodology.
- New ways of assessing learning must be evolved, to supplement or replace the present multiple-choice tests with their narrow range of objectives.

These proposals stand in sharp contrast to the practices in most classrooms, where students are passive receivers rather than active learners. Someone made the apt remark: "Math is not a spectator sport." Young people in these classrooms miss out on the sense of accomplishment that comes with solving a tough problem, the feeling of empowerment in knowing that they are capable of mastering this "hard" subject.

Settlement of Pequot Indian Claims

In his 3 May 1983 letter to the *New York Times*, Robert G. High discussed the $900,000 settlement of a claim by the Pequot Indians. Congress had approved of this settlement by a unanimous vote. The president, however, rejected it as too expensive. He proposed to pay "the worth of the land at the time of original sale—$8,091.17 for the Pequots—plus interest." Little did he realize what his proposal implied. At a rate of only 5 percent, compounded quarterly from 1856, the date of the sale, the sum would be about $4.5 million, five times the amount granted by Congress.

It would seem that the Pequots were getting the short end of the stick. No wonder Congress approved unanimously! To quote Mr. High: "The Pequots might be well advised to take the President up on his generous offer."

I had the good fortune to have taught in a New York State district that was known nationwide for desegregating its schools in 1951, three years before the historic Supreme Court decision. Our mathematics department made every effort to encourage students to continue in mathematics beyond the one year required by the state. In fact, the department chair dismissed several probationary teachers because their racist attitudes turned students away. We developed our own mathematics curriculum for those students who either did not qualify for or did not choose to follow the prescribed college-bound sequence, although they could transfer at any time to the academic courses. In the twelfth-year course we concentrated on graphing, probability, and statistics, often using newspaper articles and government data as a source of problems. This gave students the opportunity to review such necessary topics as decimals and percentages in a setting that respected their intelligence, and at the same time to discover the relevance of mathematics to the real world.

One of the most successful series of lessons involved an analysis of data for the five census areas included in our school district. First I drew a map of the district, showing the census-based sub-

The abacus helps children to understand grouping and place value. This Japanese abacus, called the *soroban*, shows 27,091. (Photograph by Sam Zaslavsky.)

divisions, and made a copy of the relevant data for each subdivision—average income, educational level, employment, housing, etc. All the students were given copies of all the materials. Then they worked in small groups, each group analyzing different aspects of the quality of life in the five areas. As they constructed and compared their tables and graphs, the contrast between the predominantly white middle-class areas and the low-income and predominantly African-American sections became stark and vivid. The students' semiconscious awareness of the injustice of the situation was brought out into the open. Lively discussions ensued, not only in math class but in their social studies classes as well. Some students were stimulated to do similar research on other topics.[13]

Almost any topic of current interest lends itself to mathematical treatment. Many issues are of both local and world significance and can be adapted to any level, from elementary to college—population growth, preservation of the environment, sources of energy, health care, taxation, expenditures for the military and for human services, and the comparative well-being of the different strata of our population and of the peoples of the world. Not only do these topics involve several branches of mathematics—computation, graphing, statistics, probability, geometry, measurement, and the use of technology, as well as the calculus at higher grade levels—but they also integrate mathematics with other subject areas and deepen students' awareness of the world around them.[14]

"Increase" Means "More"—But How Much?

My friend, a very competent teacher of high school French, asked me to give her a lesson in percentages. She knew how to compute students' test grades, a procedure she had memorized years before, but she really didn't understand the concepts that lay behind the procedure.

The writer of the following passage in *Senior Citizen News* (August 1992, page 8) may have had similar difficulties with percentages:

> In 1981, the Medicare Part A hospital deductible was \$204. Today it is \$652—an increase of over 300 percent. The Part B premium was eleven dollars a month in 1981. Today it is \$31.80—again a 300 percent increase.

What's wrong with these figures? Let's first consider the numbers for Part A: \$204 in 1981 and \$652 in 1992. The *amount* of the increase was over \$400. Round the numbers to hundreds for convenience. Comparing the \$400 increase with the original amount of \$200, we see that the *increase* is two times the original amount. This translates into a 200 percent increase, not the increase of 300 percent stated in the article.

A similar analysis of the Part B premiums also leads to a 200 percent increase, in round numbers.

Perhaps the writer meant to phrase the comparison differently. It would have been correct to say: "Medicare premiums today are three times the premiums in 1981." Using percentages, a correct statement would be: "Medicare premiums today are about 300 percent of the premiums in 1981." Think of one whole anything as being 100 percent; then three whole things are 300 percent.

People's Math

"Ungracious and uncaring" is how the author of one autobiography characterized mathematics as she had experienced it. The Brazilian historian and philosopher of mathematics Dr. Ubiratan D'Ambrosio agrees with her. He wrote: "I share the feelings of Gustave Flaubert when he says, 'Mathematics dries up the emotions.' The general public feels this way, and it is sad. We need more emotions, more passion, more love in the world. Mathematics is not against this, has not been throughout history, but after the seventeenth and eighteenth centuries it became the champion of an inhumane rationalism. I am against this."[15]

D'Ambrosio and others are looking at mathematics from a different perspective, from the cultural point of view. This new field is called "ethnomathematics," the study of the mathematics developed by groups of people in the course of their work or during other aspects of their lives. All cultures develop mathematical ideas in accordance with their needs and interests. No branch of mathematics can be more "applied" to real life, and anyone can contribute to this field.

Dr. Paulus Gerdes, a mathematician in Mozambique, speaks of "hidden" or "frozen" mathematics. "The artisan who discovered [a new] production technique *did* mathematics, *developed* mathematics, was thinking mathematically." One task of the ethnomathematician is to unfreeze this frozen mathematics, to rediscover the hidden mathematics in the pratices of a particular culture.[16]

It is generally agreed that societies in Asia, Africa, and the Americas developed mathematical ideas independent of and often earlier than western Europe by hundreds or thousands of years, and laid the basis for developments in western Europe, yet these contributions have often been ignored. Recent scholarship is attempting to fill in the gaps, to overcome the Eurocentric focus, an important area of work for ethnomathematicians.[17]

Some educators and publishers are beginning to incorporate cultural and historical perspectives into the mathematics curriculum, as Judy Richards did with African culture in her third- and fourth-grade classroom. Math comes alive as students learn that mathematical practices arose out of the real needs and interests of all societies. Such practices include

Africans and Mental Arithmetic

The Dutch trader William Bosman wrote at the beginning of the eighteenth century about West African traders: "They are so accurately quick in their Merchandise Accompts, that they easily reckon as justly and as quick in their Heads alone, as we with the assistance of Pen and Ink, though the Summ amounts to several Thousands; which makes it very easie to Trade with them."[1]

Thomas Fuller (1710–1790), known as the African Calculator, was brought to America as a slave at the age of fourteen. He may have come from a family of traders, for he was amazingly adept at mental arithmetic. For example, he could multiply two nine-digit numbers. Although he could neither read nor write and was forbidden access to any kind of schooling, as were all slaves, his reputation was such that antislavery advocates cited him as evidence of the high mental capacity of Africans. One man asked him the age in seconds of a person who was 70 years, 17 days, and 12 hours of age. Fuller gave the answer in one and a half minutes. Meanwhile his questioner was working it out with pencil and paper. When this man came up with a different answer, Fuller reminded him that he had neglected the leap years. It turned out that Fuller's calculation was correct.[2]

---

[1] William Bosman, *A New and Accurate Description of the Coast of Guinea* (1704) (New York: Barnes and Noble, 1967), 352, quoted in Claudia Zaslavsky, *Africa Counts: Number and Pattern in African Culture* (New York: Lawrence Hill Books, 1979), 289.

[2] John Fauvel and Paulus Gerdes, "African Slave and Calculating Prodigy," *Historia Mathematica* 17 (1990): 141–151.

systems of number and measurement, patterns in art and architecture, games of skill and games of chance. Students have the opportunity to learn about mathematical contributions of women and of non-European societies, a generally neglected area of mathematics. They can take pride in their own heritage, and at the same time become familiar with and learn to respect the cultures of other societies.[18]

Certainly the cultural diversity of the American people rivals any in the world. Yet we ignore the math that various groups have developed and the way that they process mathematical information. A New York City elementary school teacher, in an effort to jolt educators into providing more relevant curriculum materials, wrote a paper on black children's exposure to number concepts through the illegal numbers game. She described the system of hand gestures used to denote the numbers, and added, "When numbers are not legal, these gestures are used as double-talk around the cops." People with scant education develop the skill of remembering numbers for years. The concept of probability is recognized, as well as an awareness of odd and even numbers. Data are collected and recorded by pictures or by symbols. She concludes: "Certainly if teachers could in some way use some of this number logic with children, or at least recognize their familiarity with numbers when they meet them, perhaps number games based on Playing the Numbers might be substituted for the boring activities now presented to children."[19]

Understanding and connecting to the students' environment, providing an interesting context, is an important aspect of teaching mathematics, according to Dr. Gloria Gilmer, president of the International Study Group on Ethnomathematics. "The context engages them, and when they are engaged, they think. . . . We have a lot of sterile problems, like 'add these monomials.' Well, [the students] get all kinds of weird answers on that because it doesn't mean anything to them. To me, mathematics is extraordinarily flexible. The context of the problem can always be shifted. Mathematics interfaces with ordinary life in so many ways, we don't have to be stilted in formulating problems for students."[20]

Feminine interests are receiving attention in some programs. In the Maths in Work Project, based in London, Mary Harris utilizes the mathematics inherent in traditional women's work with textiles. For example, she compares the mathematical aspects of a design for a right-angled cylindrical pipe in a chemical factory with the mathematics of knitting the heel of a sock, and raises the question: Why is the industrial problem considered real mathematics, while the knitting problem is not taken seriously as valid mathematics?[21]

Symmetry is a feature of women's work with textiles—

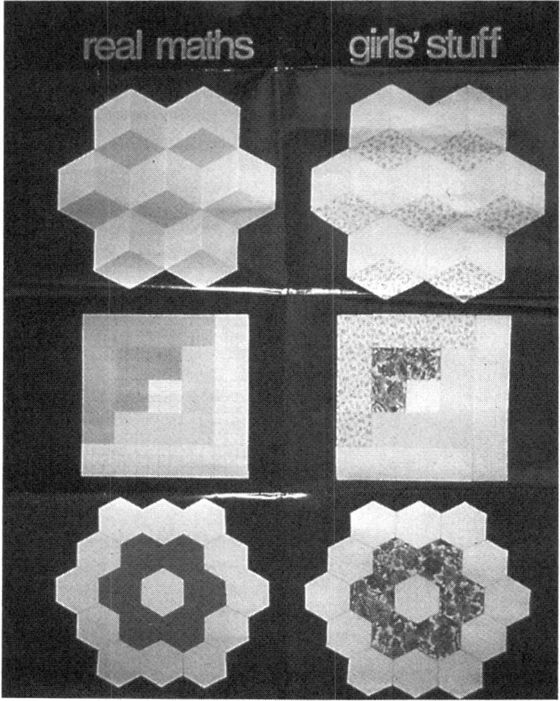

Maths in Work poster, comparing the "real maths" of abstract geometry with the "girls' stuff" of quilting patterns.
(Courtesy of Mary Harris. Photograph by Sam Zaslavsky.)

repeated patterns in knitting, quilting, weaving, embroidery. The broadened mathematics curriculum includes the study of symmetry, with applications to the arts of many cultures.[22] But I often wonder whether girls in our society are interested in this type of so-called "women's work." Perhaps applications to baseball and karate would be more appropriate!

At Clark Atlanta University, a historically black institution in Atlanta, the chair of the mathematics department, Dr. Abdulalim Shabazz, weaves culture and history into his teaching. He wants his students to "know that in their past, in their history, in their heritage, they have achieved greatness in the mathematical sciences." For example, he points to the fact that

Symmetrical patterns in Hungarian embroidery. Two articles on symmetry in Hungarian folk needlework were published in the *Journal of Chemical Education* (1984 and 1985). Symmetry is symmetry, whether in art or in molecular structures!
(Photograph by Sam Zaslavsky.)

Pythagoras and other Greek mathematicians had studied in Egypt and built upon this foundation. He has had extraordinary success in developing mathematicians; it is estimated that approximately half of the African-American mathematicians in the United States were either his students or students of his students.[23]

In a determined effort to increase the participation of Native Americans in math-related fields, a culturally based curriculum for elementary and secondary grades has been developed. The activities, appropriate for many ethnic groups of American Indians, are designed to supplement the regular curriculum and to be used in classrooms that include both Native American and other students. The program includes teacher training and the involvement of parents, with a view to reducing anxiety and emphasizing the importance of mathematics in today's society.[24]

A number of programs have been developed with the goal of bringing more young people of color into mathematics and science. (See the "Resources" at the back of book for descriptions of some of these programs.) Some have introduced multicultural aspects, as, for example, the TexPREP program, headed by Dr. Manuel Berriozabal, professor of mathematics at the University of Texas at San Antonio. Most of his students are Hispanic, and he brings in references to the Aztecs and the Maya, "to give students an appreciation of the mathematics of other cultures. . . . But, as far as becoming a mathematician, there is nothing that substitutes for hard work and achievement."[25]

I will close this chapter about school math with a modern fairy tale, a true story that took place in New Hampshire in the 1930s. L. P. Benezet, superintendent of schools in the city of Manchester, was convinced that formal teaching of arithmetic occurred much too early in the curriculum. Over a period of several years he instituted in some classes an experimental program in which there was no formal teaching of mathematics in the first five years of schooling. Rather than exposing children to a specific curriculum, teachers had them deal with numbers as they came up in the course of various activities. They handled toy money while making change, carrying out all the operations mentally. They estimated heights, lengths, areas, and distances, and checked their estimates by taking measurements. At the same time, students were getting a great deal of practice in reading, reasoning, and reciting. By the end of sixth grade the experimental students surpassed the traditional classes in standard arithmetic tests.

In one year, during sixth grade, these students had mastered a curriculum on which others had spent six years, but with poorer results. Moreover, they had a stronger understanding of mathematical concepts and could explain them with ease. These experimental students were not children of the well educated. No, they came from working-class homes where not one parent in ten spoke English as their first language![26]

Dr. Hassler Whitney liked to tell this modern fairy tale. As he often said to people who claimed they couldn't do math, "You never had a chance to see or do real math, which is easy and fun."[27] Surely the Manchester children would agree.

CHAPTER SEVEN

Everybody Can Do Math: Solving the Problem

Three years ago I finally had an opportunity to devote my time in a meaningful and productive way to learn the mathematics that I'd never studied in high school. I know I could have taken the Algebra-Geometry-Algebra sequence at any night school, but I didn't want computation; I wanted understanding, connections, *knowing*.

The Interactive Mathematics Project began offering the first year of a three-year curriculum to replace the traditional high school sequence. Its focus is on concepts, with Algebra-Geometry-Algebra being used as tools and as only part of the repertoire which can be called upon to solve and think about problems. June 1992 will mark the end of three years of participating in a high school mathematics curriculum of the highest quality. I have learned a tremendous amount, and I am not done yet!

The experience of *doing* and *thinking* mathematically has provided me with skills which transcend the discipline of mathematics. It has also resulted in my seeking out and enjoying mathematical challenges. (Who would ever have thought I could or would say *that*?!) The more you do, the more you learn.

I can't resist an analogy about overcoming fear and dislike of math. When I was pregnant, my husband and I chose to join Lamaze classes. My expectation from these classes was not that they would mean the elimination of pain. Rather, knowing as much as I could ahead of time would affect my fear of the unknown. Minimizing the fear would allow me to control the pain.

—Mary Jo Cittadino, Mathematics Educator, EQUALS, University of California at Berkeley

Everybody Uses Math

"I'm just a housewife" is a frequent excuse for avoiding math. But the woman who considers herself "just a housewife" is using math all the time. She measures ingredients, changes recipes to accommodate the number of people to be fed, times her cooking and baking, computes the cost of her purchases and checks her change, and budgets her expenditures, either formally or informally. Like the majority of women today, she may also hold an outside job that requires her to plan her time down to the minute to take care of the millions of tasks that burden her busy life.

"Just a housewife" uses geometric as well as arithmetic concepts. She knows that potatoes cook much more quickly when cut into small cubes than when left whole, although she may be unable to prove a theorem relating the surface area and the volume of a solid body. She probably doesn't think consciously about similar shapes when she enlarges a dress pattern to fit a growing daughter. She may use a good approximation to a yard length, as my mother did when she stretched the material from the fingertips of her extended arm just past her nose, although she was unaware of the standardization of the yard many centuries ago as the distance from the nose to the tip of the outstretched arm of an English king. And the measure that we still call a "foot" was originally anybody's foot, before there was a need for standardization.

The Palestinian mathematics instructor Munir Fasheh writes eloquently of his mother's mathematical knowledge: "My illiterate mother routinely took rectangles of fabric and, with few measurements and no patterns, cut them and turned them into beautiful, perfectly fitted clothing for people. . . . [Mathematics] for her was basic to the operation of her understanding. In addition, mistakes in her work entailed practical consequences completely different from mistakes in my math. . . . Mathematics was integrated into her world as it never was into mine."[1]

In a study to determine how people use arithmetic in shopping, twenty-five adults, mainly women, talked into a tape recorder while shopping for groceries in the supermarket. As they compared prices and quantities, they employed good problem-solving strategies and mental arithmetic. To corroborate their performance in the supermarket, the shoppers participated in simulated buying situations in their own homes. Working with actual prices and quantities, they had to determine which among several choices were best buys.

But these very same people, confronted with a pencil-and-paper "school-type" test involving the same kind of problems,

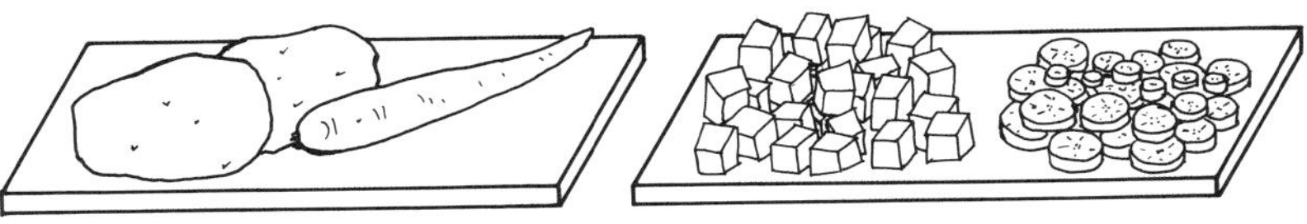

Not much surface area Lots of surface area
Both have the same volume

not only made many more errors, but reverted to school-like behavior! For instance, they talked about not cheating, and asked questions such as: "May I rewrite problems?" Although they rated over 90 percent, on the average, in the supermarket situations, their achievement level was only 59 percent on the paper-and-pencil tests.[2]

Giovannae Dennis, one of the few female African-American physicists and engineers, got her start in math and science right in her own childhood home. Even before she started school her mother showed her how to work with and alter dress patterns, and her father encouraged her to help him build and fix things. But boring school math and science turned her off, and she was placed on probation in fourth, fifth, and sixth grades. Finally her mother, in desperation, bribed her to earn A's. Eventually the bribes petered out, but by that time she was sold on math and science. Now she designs and fabricates solid-state microwave devices for IBM.[3]

Susan (age 30) discovered that she really knew more than she had given herself credit for. Returning to college after many years, she brought her math phobia with her. "At first the same fears were there. I told myself this is crazy—you know how to do this. I changed my attitude toward it. I found myself really enjoying doing problems." Susan attributes her turnaround to her maturity and to the support and cooperation of an understanding friend.

Susan's experience bears out the results of research with working people returning to college for further study. Interviews with 4,500 adult learners at Empire State College in New York State revealed that the women students, now in their upper twenties and beyond, had more confidence in their academic ability than had their male counterparts, and went on to graduate in greater numbers.[4]

We must give ourselves credit for the mathematics that we

Two Plus Two or Why Indians Flunk
by Beverly Slapin

All right, class, let's see who knows what two plus two is. Yes, Doris?

I have a question. Two plus two what?

Two plus two anything.

I don't understand.

OK, Doris, I'll explain it to you. You have two apples and you get two more. How many do you have?

Where would I get two more?

From a tree.

Why would I pick two apples if I already have two?

Never mind, you have two apples and someone gives you two more.

Why would someone give me two more, if she could give them to someone who's hungry?

Doris, it's just an example.

An example of what?

Let's try again—you have two apples and you find two more. Now how many do you have?

Who lost them?

YOU HAVE TWO PLUS TWO APPLES!!!! HOW MANY DO YOU HAVE ALL TOGETHER????

Well, if I ate one, and gave away the other three, I'd have none left, but I could always get some more if I got hungry from that tree you were talking about before.

Doris, this is your last chance—you have two, uh, buffalo, and you get two more. Now how many do you have?

It depends. How many are cows and how many are bulls, and is any of the cows pregnant?

It's hopeless! You Indians have absolutely no grasp of abstractions!

Huh?

(Reprinted with permission from *Through Indian Eyes: The Native Experience in Books for Children,* ed. Beverly Slapin and Doris Seale [Philadelphia, Penn.: New Society Publishers, 1992].)

The Language of Mathematics

Mathematics and ordinary conversation often use the same words but may have different meanings. It's one thing to be able to carry out various kinds of computations. With calculators and computers the task can be easy. It's another thing to know how to work with the data. How are the different numbers related to one another? What are we trying to find out? Understanding the language is crucial to working out the problem.

Several different expressions may indicate the same operation. For example, we can read the expression: 3 + 5, as:

- Add three and five
- Three plus five
- Three increased by five
- The sum of three and five
- Combine three and five
- Five more than three

If your home language is not academic United States English, or you attended school in another country, the difficulties are compounded. Words and phrases that have certain meanings in English may have opposite meanings in other languages. You may have learned a perfectly valid procedure or format for an operation, only to be told: "That's not the way we do it here."

For example, all of the following expressions indicate the same operation:

$$24 \div 6 \quad \text{or} \quad 24/6 \quad \text{or} \quad 6\overline{)24}$$

In English the teacher might say: "Let's divide 6 into 24" or "Let's divide 24 by 6" or "How many times does 6 go into 24?" or "How many sixes in 24?"

In Spanish the teacher might say: "Vamos a dividir 24 entre 6." The literal translation is "Let's divide 24 into 6," which is understood to mean: "Let's divide 24 into 6 parts." The Spanish expression reverses the order of the two numbers in the common English expression: "Divide 6 into 24."

No wonder people are confused. Even English speakers often get it wrong!

do know, rather than feeling "dumb" about the mathematics we feel we should know. We all discover mathematical ideas, have mathematical experiences, and formulate opinions about aspects of mathematics. This statement, that everyone can do math, doesn't imply that math must necessarily be learned in school, with textbooks and formal teaching. Most of the math that the average person uses today grew out of people's experiences over a period of many centuries. Some inventors and discoverers were not even literate, yet they used math to count, to measure time and space, to construct, to exchange goods, to design beautiful patterns, and to play mathematical games.[5]

We learn best when we can integrate the new information with knowledge we already possess. We can use our home-grown methods of problem-solving to complement the more abstract information we acquire from teachers and texts. In that way we have control over our learning; the new information is now ours. We are not merely absorbing bits and pieces that have no relationship either to one another or to our previously acquired knowledge. On the contrary, we are using this new material to construct a greater whole, enriching our understanding in the process.

Learning Styles

If I were to write my autobiography, I might have a section called "Fear of Food Processors." Some years ago we had acquired a food processor just before I left home for a week-long engagement. During my absence my husband, Sam, an engineer, figured out the workings of the machine and made good use of it. The day after my return, Susan, a professor of computer science, visited for lunch. Both Sam and Susan offered to show me how to use the processor. With an expert standing on each side of me, calling out directives—"Push that button!" "Don't put your fingers there!" "Move the bowl!"—I was soon so confused that I gave up completely. I must confess that the food processor remained unused on the shelf ever since that incident. Every once in a while I would tell myself that I will learn to use it as soon as I have finished this book (article, project, etc.). I rationalized that washing the parts and putting the thing together again is too time-consuming, or that there is no point in learning to use this old model when the newer ones are far superior, and so on.

But deep down I know that this one bad learning experience had really soured me on food processors. Finally I gave it away to a local church bazaar. Please don't follow my example. This is not the solution I recommend to math avoiders!

I understand my own learning style. When confronted with a new task or problem, I like to examine and mull over all aspects before tackling the job. With a new piece of equipment, I try to figure out the function of each part of the machine and how the parts fit together. Then I read the instruction manual, comparing the written directions with the information I have worked out on my own. Finally I am ready to use the gadget. I operate it slowly, integrating my observations with the knowledge I have already acquired. In other words, I want to take control of my own learning. I must admit that I use the telephone and drive a car with only a limited amount of information about their inner works, and when my car doesn't start, I wish I knew more.

People vary in the way they process information. Children with different types of personality traits and backgrounds come together in one classroom and are expected to conform to the instructional style of the teacher and to adapt to the methodology of the teaching materials. The resulting mismatch may have a profound effect on children's ability to learn.

Cecilia (age 18) represents such a mismatch. She was attending a summer bridge program to prepare for college when she wrote: "I always had the fear that math problems should be done quickly and I'm incapable of doing so. It usually takes me twice as long than another student to complete a problem." Evidently her teachers did not recognize her style of learning, and this failure diminished her self-confidence and ability to learn math.

Many factors enter into a consideration of personal styles in learning mathematics. Among them are:

- Preference for working alone or working in a group
- Preference for a competitive or a cooperative environment
- Preference for oral or for written methods
- Manner of tackling a new situation or problem
- Reflective style (mulling over the problem) or an impulsive style (giving the first answer that occurs to you)
- Speed of work
- Persistence in staying with a task

- Need for outside encouragement and reinforcement, or a go-it-alone attitude
- Willingness to take risks
- Spatial, numerical, logical, kinesthetic, or other approaches to problem solving

Educators now point to the need to recognize and understand cultural differences in the way people learn. The competitive, individualistic, "every person for himself or herself" style of teaching so common in our classrooms is completely at odds with the cultural norms of many ethnic groups. Children who grow up in a milieu that stresses cooperation may be turned off by a competitive classroom environment. Speed and flashcards are anathema to them. Children who are accustomed to family and group discussions to solve problems may be unable to accommodate to the teacher who expects students to sit quietly and fill in blanks on worksheets. And I believe they are right to resist this kind of instruction!

In her insightful book, *Affirming Diversity,* University of Massachusetts professor Sonia Nieto states: "Most schools favor a highly competitive and individualistic instructional mode. In this kind of environment, dominant-culture children and males are more likely to succeed, whereas students from other cultural groups and females may be at a distinct disadvantage. By combining this style with a more cooperative mode, the learning and cultural styles of all children can be respected and valued."[6]

"Is there an [American] Indian learning style?" asks Floy C. Pepper. She answers her question in these words:

> A growing body of research suggests that the child-rearing practices of European-American and Indian parents produce notably dissimilar styles of learning among their children. Sharing, cooperation, group harmony, modesty, placidity and patience are greatly prized among Indian families, while competition with others and singular notoriety are considered embarrassing and dishonorable instigators of dissonance. The European-American youth is taught to compete against others and win, to "make a name" for himself or herself. . . .
>
> The danger of stereotyping lurks beneath the determination of one learning style for all. . . . No *absolute* In-

dian behavioral learning style exists. There is a wide variety of individual differences which can be viewed as tendencies.[7]

Pepper's warning against stereotyping applies to all cultural groups, to women and to men. The styles that we are discussing are tendencies, not absolutes.

Women's Ways of Knowing Are Not for Women Only

The typically "feminine" socialization of girls in our society causes them to blame themselves when they don't achieve in school mathematics. They have been brought up to depend upon outside authority, to feel powerless and not in control of their lives. Afraid to venture their own opinions for fear of being challenged and crushed, they dutifully accept math as it is generally presented—a collection of rules and procedures devised by some kind of magic. The authority—the teacher or the textbook—is not to be questioned. Usually speed is a factor in school learning, ranging from flashcard drill to timed standardized tests. There is little opportunity for exploration, nor is thoughtfulness expected. The one right answer is the goal and rote memorization is the only way to achieve that goal. Anyhow, it all makes no sense and has no relevance to women's own lives or to the real world.

No wonder many girls lose confidence as they enter the middle grades. And confidence in their ability to learn mathematics is one of the main factors that determine whether young women will continue in math. Compared with boys who achieve on the same level, girls are far less confident of their abilities. It is important to note that these conclusions apply to the white middle-class students who have been the main subjects of research in this area of learning.[8]

My own experience with students of all ages convinces me that math as taught in school doesn't make sense to many boys. It's not only girls who find it hard and unappealing. But white middle-class boys are more likely than girls to have career goals that include a background in mathematics. Encouraged by the expectations of their parents and teachers, and with the future in mind, they are more willing to stick it out.

The authors of several recent books have explored women's

Mysterious Arithmetic

In 1879 the U.S. Training and Industrial School in Carlisle, Pennsylvania, opened its doors to American Indian students. The goal was to separate them from their people and turn them into "white" people. They were forbidden to wear traditional dress, to practice their customs, or to speak their own languages.

Sun Elk, from the Taos Pueblo in New Mexico, tells about his seven years at Carlisle. "They told us that Indian ways were bad. They said we must get civilized. It means 'be like a white man.' "

Part of "getting civilized" was learning arithmetic. As Sun Elk tells it:

> It was lessons: how to add and take away, and much strange business like you have crossword puzzles only with numbers. The teachers were very solemn and made a great fuss if we did not get the puzzles right.
>
> There was something called Greatest Common Denominator. I remember the name but I never knew it—what it meant. When the teachers asked me I would guess, but I always guessed wrong. We studied little things—fractions. I remember that word too. It is like one half of an apple. And there were immoral fractions.
>
> —Peter Nabikov, ed., *Native American Testimony: A Chronicle of Indian-White Relations from Prophesy to the Present, 1492–1992* (New York: Viking, 1991), 222.

No doubt, to the schoolmasters' way of thinking, only a thin line separated "improper" from "immoral." I wonder, are children still being taught "proper" and "improper" fractions? On a deeper level, of what value was Sun Elk's out-of-context "education" in arithmetic?

ways of learning. Two aspects stand out: cooperation and context. Women find it emotionally satisfying to work with others, a style that is discouraged in most classrooms. Secondly, a woman wants to know the context of any particular problem or procedure, how it relates to what she already knows, how it fits into her world. These authors categorize styles of reasoning as "separate," referring to traditional abstract ways, and "connected" or "relational," the style more typical of women.[9]

Ellen (age 20) expresses her longing for meaning in mathematics. "I have now reached the point where I no longer enjoy math. I spend hours working problems to come up with a wrong answer. Or if the answer is right I have no idea what it means. I feel that 'learning' this information is a waste of my time because we are not taught how to generate the problem which we are solving."

The way math is generally taught in school favors the masculine mode of learning, say instructors who have worked with math-fearing adult women. Men of the dominant culture are socialized to be individualistic and competitive. Their upbringing, their experiences in and out of school, accustom them to deal with ideas in the abstract, not necessarily in a relevant context. Men talk more than women, claims Deborah Tannen, author of *You Just Don't Understand: Women and Men in Conversation,* and are eager to exhibit their skills in argument, to prove themselves right—and the other person wrong. Women are more likely to look for positive aspects, for areas of agreement with their colleagues, and are reluctant to be challenged publicly.[10]

Does a competitive, rather than a cooperative, atmosphere motivate people to learn better? In a competition a few win, but most people lose. In her best-selling book about self-esteem, *Revolution from Within,* feminist author Gloria Steinem refers to psychologists' studies and concludes: "In fact, a competitive system perpetuates itself by keeping self-esteem low and making even the winners constantly needy of more success."[11]

Educators investigating the affective aspects of learning find that working in collaborative groups improves the learning of all students. As for context, people naturally relate best to that which is most meaningful, whether in terms of what they already know, or their environmental or cultural background.

In sum, women's ways of knowing are the preferred ways of most people.[12]

Dr. Dorothy Buerk, a mathematics professor at Ithaca College, has applied ideas about women's ways of knowing to the learning of mathematics.[13] Adult math-avoidant students, both men and women, have profited from her insights. Among the principles she enumerates are:

- Experience the problem, relate it to your personal world, clarify the language so that it makes sense to you.
- Rely on your intuition and feelings; use a contextual mode of thinking.
- Look at the limitations of any solution to a problem and the conflicts that remain.
- Make exceptions to the rules when you think it is appropriate.
- Be reluctant to judge.

Buerk suggests that mathematicians actually do work on the basis of these principles when they develop their mathematical ideas. It is mathematics *instruction* that is at fault by giving an entirely different impression about the field of mathematics. Students don't realize that mathematicians often flounder in their search for a solution, that they may work with many concrete examples before they can generalize, and that they can experience great joy, as well as dismal depression, in the course of their work.

Cooperative learning is one key to the success of the Mathematics Workshop Program at the University of California at Berkeley. Mathematician Uri Treisman set up this program to deal with the poor performance of many African-American and Latino students in the difficult first-year calculus course. Although they had been outstanding students in their high schools, they were unable to cope with the stiff academic demands at the same time that they made the multiple adjustments to college life that face minority students in predominantly white institutions.

First Treisman investigated how the African-American freshmen studied. They worked very hard, but they worked alone. In fact, they refused to ask for help when they were unable to solve the assigned problems, fearing that it might be a reflection on their ability, and thus reinforce the negative stereotype that minority students had been admitted inappropriately. In contrast,

An Unsolved Problem

One of the myths about mathematics is that math is complete, that everything has already been discovered. All we can do is study what others have done.

To disprove that false idea, I will give you an example of a problem that has not been solved, although people have worked on it for over two and a half centuries. The nice part is that you don't need more than elementary school arithmetic to understand the problem.

In the 1700s a mathematician named Goldbach stated his theory:

> "Every even number larger than two can be expressed as the sum of two prime numbers."

A prime number is a number that is divisible (without remainder) only by one and by the number itself. The first few prime numbers are 2, 3, 5, 7, 11, 13, 17, . . .

> Examples of Goldbach's theory: 8 = 3 + 5; 10 = 3 + 7 or 5 + 5.

The number 10 can be expressed as the sum of two prime numbers in two ways.

No one has been able to prove that Goldbach's theory, called "Goldbach's conjecture," is either true or false. It has been true for every even number ever tested. Since there is an infinite number of even numbers, it is not possible to test *all* even numbers. That's the big difficulty.

Now, if someone were to discover just *one* even number that cannot be expressed as the sum of two prime numbers, we would know that the conjecture is false. Only one *counterexample* is needed to prove a theory false. To date, the conjecture has been neither proved nor disproved.

Asian students would first work on the assignment alone, then get together in "study gangs" to criticize each other's solutions and work on the difficult problems. These groups also served as centers of social life and for mutual help with all types of problems. But when Treisman tried to set up remedial groups for the black students, they rejected the proposal as unnecessary.

After considerable thought, Treisman developed the Mathematics Workshop, a program that stresses academic excellence and supports growth of a student community around shared academic interests. Students work individually on challenging mathematical problems—problems that are even more difficult than those in the regular assignment. After they have done their best work alone, they join other students for further exploration. A faculty person is called upon only when no one in the group is successful.

The goal of the Mathematics Workshop, a project of the Professional Development Program, is to foster leadership skills and promote academic excellence in minority students. In the mathematics sessions, participants are encouraged to discuss all aspects of the problem, not merely to look for answers. The program has been outstanding in developing independent learners who complete college and go on to successful careers or graduate study in the sciences and mathematics. By 1992 over one hundred universities nationwide had adopted this program.[14]

You Can Overcome

Many of the people attending college today are not like the traditional students of the past. Some are considerably older. Some are victims of gender, race, and class biases that have affected them both academically and psychologically. These "nontraditional" learners may not be prepared for the usual college math courses. Most colleges have set up remedial courses and tutoring laboratories to help them to make up the work they should have mastered in precollege institutions. For various reasons, these measures are not always successful. Recall Alberta's comment in chapter 1: "The tutor that I had made me feel dumb, so I never returned. . . . I ended up failing the course."

Take heart from Marcia's experience. A student at the same college as Alberta, she was more fortunate in her choice of instructors and tutors. As a child in Guyana she had always feared and hated math "with a passion." After flunking out of one college because of failing math grades, she found success at last with a patient instructor and tutoring four times a week.

Alberta and Marcia should have been Mary's friends. Mary (age 19) describes herself as having a "gifted talent for mathematics. I inherited that trait or talent from my mother." In high school she held study sessions with her friends. "First they had to open up to me and let me know what caused them to be afraid of math. For many of them fear was caused by themselves because they psyched themselves into believing that math was something that could not be accomplished."

Many instructors who have worked with college students in a variety of settings—college-level mathematics courses, developmental or remedial courses, math labs, and counseling clinics or workshops—have developed successful methods to help people overcome fear and avoidance of mathematics. Their students ranged in age from recent high school graduates (or dropouts in some cases) to senior citizens. The programs in which these students were enrolled varied from the two-year college associate degree to graduate school. Some students, long absent from the academic scene, needed to polish their skills and to rebuild confidence in their ability to do mathematics. For others, unpleasant experiences in their early school years had convinced them that they were "dumb" in math. People with inadequate math backgrounds for the careers of their choice constituted another group, people who were behind their fellow students either because they had not chosen earlier to take the necessary courses, or because of societal conditions beyond their control. Some of the people involved in these studies were math avoiders, others were math fearers, and still others were rusty in math. Many people suffered from all three conditions. Fear of math had led to math avoidance, and, consequently, to the loss of some of the skills they had once possessed.

The following recommendations have helped many of these people.[15]

Review Past Experiences

It seems that most of the people with whom I talked while gathering material for this book have suffered in some way from experiences that undermined their confidence in their ability to do math. A helpful procedure in bolstering confidence is to review the experiences that led to this condition, to talk about them with other people, and to elicit *their* experiences. It's so comforting to know that you are not alone!

Examine your feelings about mathematics. If you dislike or fear the subject, try to pinpoint the incident that first turned you away from math. Perhaps it all started with a humiliating experience in an elementary school classroom—failure to understand a particular topic or stumbling over the multiplication tables. How did you feel when you had to take a math test? Often the physical symptoms of anxiety—headaches, stomach pains, diarrhea—persist into adulthood in such severe form as to prevent the victim from learning at all. Some clinics offer relaxation exercises to help students overcome these intense emotional reactions to mathematics. Ruedy and Nirenberg recommend that you tell yourself many times a day and write on all your memos: "I am brilliant!"[16]

"For most members of the public, their lasting memories of school mathematics are unpleasant—since so often the last mathematics course they took convinced them to take no more," states the report *Everybody Counts*.[17]

This was certainly true of Molly (age 28). "I had my first difficulty with math in intermediate algebra. I didn't get help from teachers, which put me behind. As a result my confidence level dropped way down. I felt inadequate and therefore was embarrassed to go for help. Intermediate algebra was the last math class I took in high school." Years later she enrolled in a college course which was "basically a replay of high school algebra. I didn't go for help when I needed it and procrastinated on my assignments. I got halfway through the course and withdrew." Eventually she learned "that it is not unusual for math students to have to work very hard at understanding concepts. . . . Now I have fun at math. It gets harder every class I take, but I really do believe that I can learn the course content if I am willing to expend enough effort to learn."

Greta (age 31), who grew up in Germany, had a similar experience. "I hated math and I hated school in general. . . . I don't even know how I went through the last grades and I remember nothing about math. . . . [In college] I am forcing myself to study. I think that's the biggest problem: to get going. Once I do it, it is okay. I don't know why it is hard to get going; maybe old math memories from school."

Juanita (age 50) feels that she got through elementary and high school math by using unconventional methods, "which worked to get me to the correct answer without really solving any problems," or by copying from her classmates. In college, "I swore I would never do another math course as long as I lived. I was faithful to this promise until I decided to pursue graduate work and was required to take statistics." She passed with the help of a tutor. "I am currently a doctoral candidate with the burden of having to keep a tutor 'on retainer.' But my lack of math background has cost me more than just money. My self-esteem is the pits when I think of my deficiencies in math. I am embarrassed by my condition and only recently have I acknowledged that I must do something. I have made some efforts to overcome my anxiety and I have been improving. But I believe I will always think of myself as math illiterate."

Many mathephobics dislike math because they have formed a mistaken view of what mathematics is all about, and their school courses have done nothing to enlighten them. In the previous chapters are some examples of the scope of mathematics—the many useful and interesting applications and the new field of ethnomathematics. Mathematics is a creative, dynamic, growing area of thought and action, not at all the dull, dry, dreary subject that most people have been taught in school.

Joan (age 34) makes an appeal that should be heeded. "Most of the people, younger or older, who fear math justify their fear with anger over the thought they are learning something that is only useful to satisfy a school requirement. If some type of instruction or orientation was given about the validity and usefulness of math in all subjects, it would perhaps begin to turn fear into curiosity. Positive reinforcement at the early levels also seems to be a very critical factor."

Analyze Your Learning Style

Students often feel lost and powerless because of the mismatch between the style of teaching they encounter in the classroom and their own personal learning style. Analyze your style of learning and try to adapt the task at hand to that style. At the same time work on broadening your approach so that you can function successfully in many kinds of situations.

Why Not Divide by Zero?

Why don't we divide by zero? For a couple of reasons; both refer to multiplication by zero.

First let's see how multiplication and division are related to each other. For example, $6 \div 3 = 2$. We check by multiplication: $3 \times 2 = 6$.

Multiplication and division are considered *inverse* operations. One operation is the inverse of the other.

Now let's see what happens when we divide by zero.

Let's say that $6 \div 0 = N$, where N stands for some unknown number. To find the value of N, we'll refer to multiplication: $0 \times N = 6$.

Even if you haven't memorized the multiplication tables, you know that zero times any number is zero. There is *no* number that makes this statement true: $0 \times N = 6$. Therefore, $6 \div 0$ has no solution.

Be careful not to confuse division by zero with dividing zero by a non-zero number. For example, consider $0 \div 6 = N$. The corresponding multiplication problem is: $6 \times N = 0$. N must be equal to zero; no other number

(*Continued*)

It's never too early to learn that you can't divide by zero!

Why Not Divide by Zero? (*Continued*)

makes the statement true. Therefore, zero divided by a non-zero number equals zero: $0 \div 6 = 0$.

Now consider this case: zero divided by zero. Follow the same procedure.

Let's say that $0 \div 0 = K$, where K stands for some unknown number. To find the value of K, we'll set up a corresponding multiplication problem: $0 \times K = 0$. Lo and behold, K can be *any* number: 3, 7.9, -2^3, 0, etc. Zero times any number is always zero.

So we have two cases involving division by zero:

- A non-zero number divided by zero has no solution.
- Zero divided by zero has an infinite number of solutions.

Isn't math amazing?

Experience Each Problem

Reading a mathematics book is far different from reading a novel, an essay, or a history book. You will probably have to go over one section of the text several times. Have paper and pen ready. Make notes in an organized way. Write down new terms, definitions, symbols, and ideas. After you have studied the worked-out examples, close the book and try to do them yourself. Note any aspects that you don't understand and get help as soon as possible. Don't just gloss over them and hope that you will catch on sometime in the future.

The introduction of concrete math materials in elementary classrooms has helped to make math comprehensible to young children. But what about adults who never had the benefit of such concrete forms of learning? It isn't too late for them. Instructor Phyllis Steinmann uses Cuisenaire rods and Algebra Blox of her own design to investigate the properties of numbers and relationships in algebra. She sees her role as that of a facilitator in helping students to construct their own knowledge by studying and reflecting upon patterns, attributes, and relationships. By no means

Manipulating Algebra

Algebra tiles are similar to the Base Ten materials described on page 22. A set of algebra tiles includes:

- Small squares, representing *ones* or *units*
- Strips measuring one unit by x units
- Large squares measuring x units along each side

What do I mean by a measurement called "x units"? The tile labeled x represents the variable. In practice, x can be any convenient length.

Example: Use the tiles to find the product of $(x + 2)$ and $(x + 3)$.

Procedure: Form a rectangle having the dimensions $(x + 2)$ by $(x + 3)$, using as few tiles as possible.

What is the total value of the tiles in the rectangle? The rectangle contains one large square, five strips, and six units. Therefore:

$$(x + 2)(x + 3) = x^2 + 5x + 6$$

Example: Factor the expression: $x^2 + 5x + 6$.

Procedure: Combine the following tiles to form a rectangle:

One large square (x^2), five strips ($5x$), and six units (6).

The dimensions of the rectangle are $(x + 2)$ and $(x + 3)$. Therefore:

$$x^2 + 5x + 6 = (x + 2)(x + 3)$$

Negative quantities can be represented by tiles of a contrasting color.

Example: Find the product of $(x - 1)$ and $(x + 3)$.

Procedure: Form a rectangle having the dimensions $(x - 1)$ by $(x + 3)$.

The rectangle contains one large square, three positive strips and one negative strip, and three negative units. Therefore:

$$(x - 1)(x + 3) = x^2 + 2x - 3$$

(*Continued*)

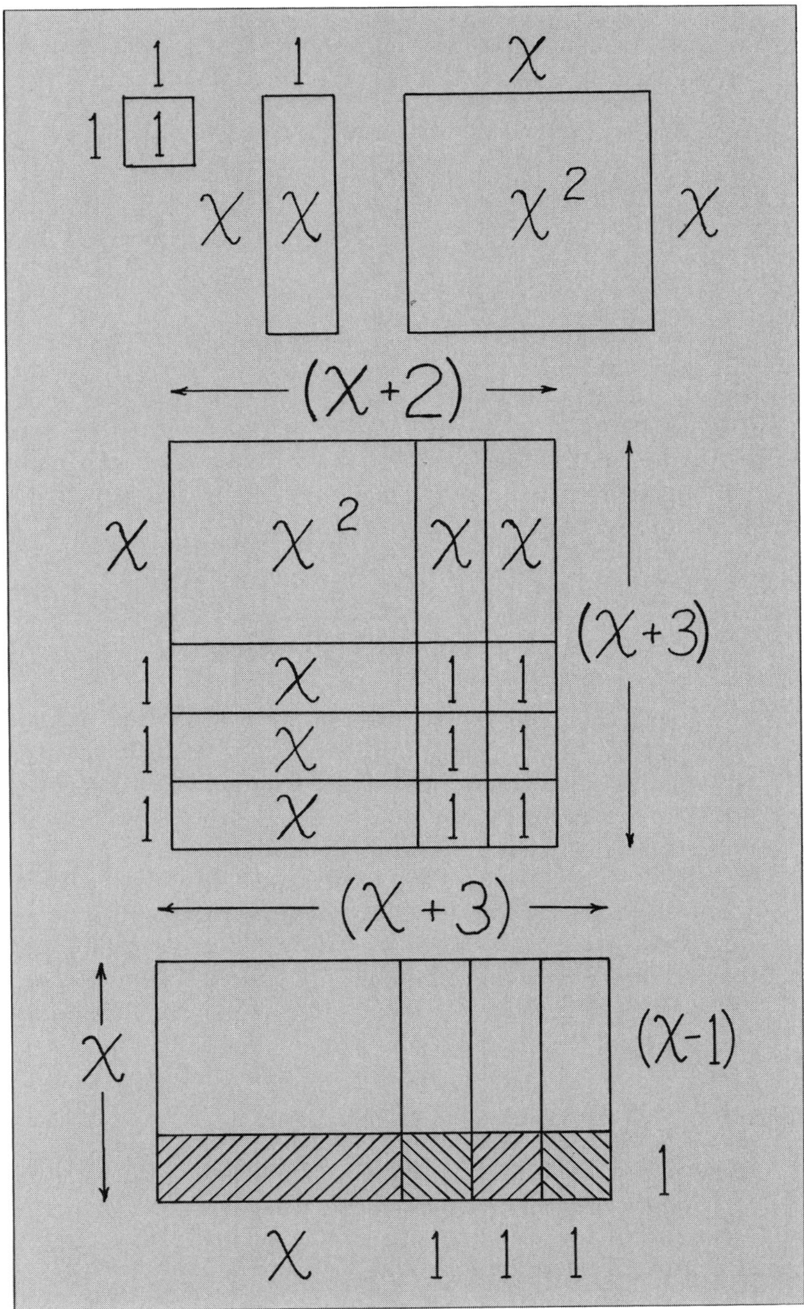
1
1
1
1
χ
χ
χ
χ²
χ
(χ+2)
χ
χ²
χ
χ
(χ+3)
1
χ
1
1
1
χ
1
1
1
χ
1
1
(χ+3)
χ
(χ-1)
1
χ
1
1
1

are concrete materials only for remedial use. Mathematicians, too, use them to elucidate knotty problems. They are particularly appealing to people who learn best by touching and feeling.[18]

Bob (age 20) would certainly have benefited from the use of concrete materials. In the lower grades, subtraction with regrouping—often called "borrowing"—was a complete mystery to him. "My real frustrations in math came in Algebra I. Algebra is and will always be one of those words I despise. Only one concept, a very important one I might add, troubled me in Albegra, and that was factoring."

The following suggestions for solving problems are based on my experiences and those of several other concerned math instructors. These suggestions may not apply to every situation. You will have to use your own judgment. But the more fully you experience each mathematical encounter, the more skilled you will become in solving problems in the future and the more you will enjoy doing mathematics.

- Simplify the language and restate the problem in your own words.
- List the given facts and the relationships among them. What are you asked to find?
- Work with the given information in several different ways—make a table, draw diagrams, use concrete materials, etc.
- Relate this problem to other mathematical ideas that are similar in some ways, and note both the similarities and the differences.
- Relate the problem to your own experience. Try to form mental images that involve the given information. Teacher and therapist Elisabeth Ruedy says: "I love teaching math by imagining, drawing, thinking in context, trial and error. In short, I love teaching it away from the dry, left-brain, step-by-step, there-is-only-one-way methodology that scares away the more creative, dreamy child or adult not so interested in precision and linear thinking (but bent on fully understanding the task at hand)."[19]
- Work out some simpler examples of the same relationships and use these examples to discover a pattern.
- Arrive at an approximate answer that seems logical.
- Hopefully, work out a solution.
- Check the solution in more than one way, if possible. Is it

Are We All Violent Criminals?

It is said that we often learn more from our errors than from our successes. Here is an opportunity to learn from someone else's mistakes.

According to an editorial in the *New York Times* (25 January 1992) entitled "Death, Life and the Presidency," in 1990 states executed 23 murderers. "These constituted less than .001 percent of all murderers that year," said the editorial.

Statements like these arouse my curiosity. How many murderers were there in 1990, I wondered. This is how my thoughts ran:

> If 23 executed murderers = .001 percent of all murderers, then (multiplying both numbers by 1000) 23,000 murderers = 1% of all murderers, and (multiplying both quantities by 100) 2,300,000 murderers = 100% of all murderers. Wow! Is it possible that 2,300,000 murderers live in our midst?

I read the next statement in the editorial: "These [23 executed murders] were only .000004 percent of all violent criminals."

I continued with my analysis:

> If 23 executed murderers = .000004% of all violent criminals, then (multiplying both quantities by 1,000,000) 23 million = 4% of all violent criminals, and (multiplying both numbers by 25) 575 million = 100% of all violent criminals. The population of the United States was about 250 million in 1990. In other words, there were over twice as many violent criminals in the country as the entire population! Something must be wrong!

I didn't look in the "Corrections" section of the newspaper the following days to see whether they had found the

(*Continued*)

Are We All Violent Criminals? *(Continued)*

errors. Who reads the corrections, anyhow? The damage had been done.

Moral: Always check to be sure that your answer is reasonable.

close to your approximation? If an answer key is available, use it. That's not cheating! If you have not been able to reach a solution, you may be able to use the answer in the key to work backward, and then put it all together.

- Think about your solution. Might you have done the problem a different way? I often wonder at myself—why did I solve it by such a roundabout route when I could have done it more efficiently in a way that would give me what mathematicians like to call an "elegant" solution. All of this experience is valuable, not a waste of time. That's how math should be done.
- Explore further. Make up similar problems. Change some of the conditions in the problem and investigate "What if not? . . ."[20]
- Even if you don't get the answer, keep all your work. Learn to find your errors and to analyze the nature of your mistakes, one of the best learning experiences. Arrange your work in a way that will enable you to go back and review your thinking processes. Use a pen, or a pencil without an eraser, so that you won't be tempted to erase what you have written before you have gone over your work and found the errors. You may find that you have wrong answers for your calculus problems simply because of careless mistakes in arithmetic or in elementary algebra. In computer programming they call it "debugging." Most computer programs have bugs in them initially, and debugging may be more time consuming than devising the original program. Computer programmers expect that to happen.
- Don't let frustration get you down. Take a rest from the problem by doing something else—another problem or a different activity. I find that taking a shower helps me to

think clearly. I wonder how many problems have been solved in the shower!

Carlos (age 18) had a tough time with mathematics for a while. Then he learned to relax. "Now I have learned not to pressure myself into learning material all at once, but to take my time and relax when I am studying."

Shelly (age 40) wrote: "Now that I'm, let's say, a mature student I realize that what math requires is time. Most people get frustrated and aren't willing to spend the time that it requires so therefore they dislike it. I think if other subjects come easy for some students and then they hit math and it doesn't come as easy, they think they don't have the aptitude for it and don't really try."

In chapter 2 I spoke about Herbert, the man who needed a better-paying job in order to adopt a child. His frustrations resulted from studying too much. "I study six or seven hours a day but it doesn't seem to do any good," he told his instructor, Phyllis Steinmann. Her response was inspired. "I want you to stop studying so many hours a day. There will be a limit of no more than two hours, and not that long if you find you are becoming frustrated." She promised him a grade of C if he agreed to her conditions. Reluctantly he went along with her. With the promise of a passing grade he relaxed enough to earn a B in the course![21]

Helene Walker was in her sixties when she earned her doctorate in mathematics after working at it for thirteen years. She wrote in her math autobiography: "If you don't see how to do a problem immediately, that doesn't mean you can't do it. You have to give it thought and give it time to jell in your mind. Let it rest a while, then go back to it—you will have a fresh perspective."

Two Or More Heads Are Better Than One

Sharing experiences with others in a small group, or even with one other person, is helpful in several ways. It's reassuring to know that you are not the only one having trouble with math. Furthermore, each person can contribute her or his viewpoint about the problem at hand, thereby leading to a richer solution and a more valuable experience for each member.

In her work with a small group of intellectually able but math-avoidant women, Dorothy Buerk first had the participants "experience" the problem individually. Afterwards they got together for a group discussion.

> When discussion did ensue, the focus was on the question rather than an answer. The women were encouraged to ask questions about the meaning of the problem, to clarify any puzzling terms, and to share the mental images that the problem brought to mind. I believe that this "experiencing" step was important since it allowed each woman to make the problem meaningful for herself and to clarify it for herself both visually and verbally. Only when every group member "saw" the problem did resolution become the focus. This approach encourages reflective and personal thinking; it provides the time to make connections, to make meaning for oneself, and put aside the ideas that may distort the problem or inhibit a solution.[22]

Writing to Do Math

More and more in recent years mathematics educators are recognizing the value of writing to do mathematics. They encourage students to keep individual journals in which they can reflect upon their feelings about the subject—a chance to get back at the hated third-grade teacher!—and record their thought processes as they attempt to solve problems. As they look back at the earlier pages, students realize how much they have progressed, rather than focusing on and getting nervous about the material yet to be learned.

If you are learning math on your own, or are enrolled in a class in which journals are not expected, you can still incorporate writing into your study procedure. One way is to divide the paper in half vertically, using one column for doing mathematics and the other for writing reflections about your thought processes and the feelings that are aroused as you try to reach a solution. Some instructors suggest three categories: solution procedure, thought processes, and feelings.

Early in her courses Buerk asks her students to write a statement that begins: "For me mathematics is most like . . ." Here are some of their thoughts:

> For me mathematics is most like the military draft: the amount of interest I have for math is comparable to how much soldiers like the thought of dying.

What is EQUALS?
It's providing a learning climate that encourages everyone.
Strato
It's becoming aware of issues, attitudes, and the need for mathematics.
It's thinking, talking, and writing about mathematics in groups of many sizes.
It's using all of the tools of mathematics: calculators, blocks, computers, diagrams, and graphs — with many problem-solving strategies.

> For me mathematics is most like a closed door: all the information is there, only I don't have the key.
>
> For me mathematics is most like quicksand: I find myself drowning in a mass of equations and variables, finding that the more I struggle, the more I drown.[23]

These people reveal how powerless they feel in the intimidating presence of mathematics.

Rutgers University professor Arthur Powell works with severely mathephobic students, victims of racial, gender, and class oppressions and of inferior inner-city schooling. Many of these students need remediation at the level of arithmetic and elementary algebra. Journal writing forms an integral part of his teaching methodology. He collects the student journals weekly and writes nonjudgmental comments about the entries to encourage further exploration.

As a first year "Developmental Mathematics" student, José collaborated with Powell in a study of writing. He also wrote a math autobiography for me. He had entered college with an inadequate background in mathematics. Junior high school math assignments had been to do "fifty problems, all of the same type." Although he had gone as far as trigonometry in high school, he "felt stupid" and "left high school feeling very frustrated with my knowledge of math." His experiences in the "interesting and challenging" Developmental Mathematics classes changed his feelings both about mathematics and his ability to learn the subject. After passing with high grades, "I now feel comfortable enough to take just about any course and feel that I can do well in it."

Facing the Unknown

Changing the way you think about math, how you do math, and what you expect of math can be a threatening experience. That fact was brought home to me many times in the course of my teaching.

One year I decided to start all my classes with a unit on finding patterns, a basic aspect of doing mathematics. Some students reacted with enthusiasm, while others were completely bewildered. "How do we do it? You didn't give us the formula!" A week later one of the students brought me a "drop" form to sign. When I asked why he was dropping the course, he replied, "I don't need more math. Besides, you're just playing games!" His conception of mathematics was so skewed that he considered *thinking* to be "just a game," just idle recreation.

My experience in teaching probability to nonacademic eleventh-grade students was similar. They were so confused by the degree of uncertainty inherent in the subject that I had to postpone the topic until the following year. In contrast, a first-grade teacher in my course for practicing teachers decided to teach her inner-city class to toss pennies and predict the outcomes. To her surprise, the kids already knew all about pitching pennies and

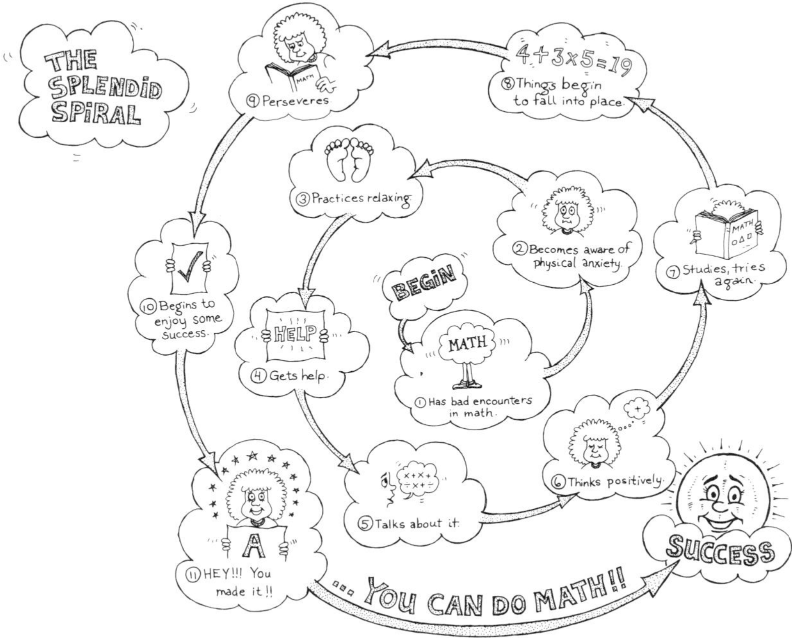

were completely absorbed in the lesson. They had not yet been exposed to dreary, dull, deadening school math with only one right procedure and one right answer for each exercise.

Mathematics is a language, with its own terms and symbols. Like a foreign language, learning math requires time and practice. Like a foreign language, the more you use it, the more skilled you become in the use of it. The ability to do math well is not a "God-given" talent, but one that requires years to develop. Some fortunate people get an early start. For those not so fortunate, persistence will bring results. Everyone can experience the thrill of the "Aha!" moment, when everything comes together in a flash of sudden understanding.

Mary Ann (age 40) wrote about her current success in college, after only one semester of high school math and a hiatus of many years: "I've been at a community college for a couple of years, and started math from the very basic review up to where I'm at now with calculus. I enjoy math even though it's very difficult and frustrating at times. . . . I think making myself do whatever it is I fear will eventually generate success. When we do things we fear sometimes our fears loom much larger than the actual experience. For myself, meeting my fear head on helps me to finally believe I can change it."

In the conclusion of their very readable book, *Where Do I Put the Decimal Point?*, Ruedy and Nirenberg express the hope that apprehension "has given way to a new assertiveness. . . . Along with the assertiveness comes a helpful irreverence toward those numbers that are used to manipulate us; inflated or skewed statistics as they are used in politics and advertising; the various numbers that people use to label us, such as grades, IQ, and medical risk factors."[24]

I have the same hope for the readers of this book!

CHAPTER EIGHT

Families, the First Teachers

I began developing an understanding of mathematics when my daughter entered first grade and expressed a dislike for math because of problems she was having in the face of timed tests around adding and subtracting sheets of computations.

Helping our child develop mathematical understanding has meant an opportunity for both my husband and me to learn—truly learn—what we had done rotely without understanding in the past. We have both told our daughter that working with her to learn about fractions has given us a second chance to learn them ourselves. We could see from the look on her face that this simple statement impacted on her in a meaningful way.

I volunteered to become prepared as a class leader in the FAMILY MATH Program when Kara was in the first grade. Despite not understanding or *knowing*, I went ahead and provided series of classes to the families in my daughter's school and, eventually, to the families in her district.

Facilitating classes has meant my having the opportunity to learn many ways of thinking, of approaching problems, from all these families who came to the classes. This includes the children as well as the adults. Each encounter has meant a deepening of my understanding.

—*Mary Jo Cittadino, Mathematics Instructor, EQUALS, University of California at Berkeley*

"Infants Have Math Ability at 5 Months, Study Shows," proclaimed a newspaper headline. According to the article, "children as young as five months can perform rudimentary addition and subtraction, indicating that humans are born with an inherent mathematical ability that is functioning well before they are taught arithmetic."[1]

As I read this article, I couldn't help contrasting this revelation with my children's school experiences several decades ago. In first grade they were taught number combinations up to a sum of ten—or was it five? Parents were told not to teach their children to read before they entered school, for fear that they would be bored with the school curriculum. We were not warned not to teach them arithmetic; that subject was not considered important. Somehow my sons did learn to read earlier than the school authorities permitted them to acquire that skill. As a result, one son was punished by being deprived of his reader for the rest of the year because he had almost finished the book while the class was supposed to be on page 10. I managed to have him transferred to another class, but it turned out to be not much better.

Today we know a great deal more about how children learn. I'd like to believe that the authoritarian teachers of several decades ago have since retired and been replaced by people well versed in the psychology of learning, but I know that is a utopian dream. Ultimately, it is the parent, the family, or the caretaker, working with the school, who must take responsibility for their children's learning.

Success Stories

When I look back at the early experiences that influenced my choice of career, I appreciate the importance of informal mathematics in my childhood. I still recall the excitement of being allowed, as a preschooler, to join my older brother and cousins in card games like "Steal the Bundle." Not only did I reinforce my skill in counting and reading numbers, but I learned to make number combinations so that I could more easily "steal" another player's stack of cards.

By the time I entered fourth grade I had amassed years of experience as a salesperson and cashier. Until I was four years old my parents operated a small stationery store. I was allowed to sell candy, three-for-a-cent and four-for-a-cent, to the local youngsters. Watching my father keep accounts in a notebook fascinated me, and I, too, had to have my own notebook in which I wrote columns of figures. Before long I could write all the numerals from one to 100.

When I was seven my parents opened a clothing store. We all worked there—Mom, Dad, and I. Besides measuring yards of ribbon and determining the correct stocking size by wrapping the stocking foot around the child's fist, I was entrusted with the operation of the cash register. Dad taught me to "count on." To

Steal the Bundle

Here are the rules, as I recall them, for the card game we called "Steal the Bundle." There is nothing hard and fast about these rules. In fact, I will suggest alternatives. You may want to make up your own rules, or ask your child to suggest variations.

Materials: One standard deck of 52 cards. You may want to start young children with fewer cards—for example, four cards each of One (Ace) to Seven.

Players: Two or more.

To Start: Deal four cards to each player, one at a time, face down. On each round place one card on the table, face up, for a total of four.

Rules of Play:

Easy: Each player in turn tries to match a card in her hand with a card on the table. Assume she has a 7 in her hand, and she sees a 7 on the table. She makes a pair of 7's and places them, face up, in front of her. This stack of cards is her "bundle," and she will build it up as she plays. If more than one 7 lies on the table, she

(*Continued*)

Steal the Bundle (*Continued*)

may pick up all the 7's with the 7 in her hand. Alternatively, assume a 2 and a 5 lie on the table. She may pick up the 2 and the 5 with her 7, placing 7 on the top of her bundle. She may also pick up any 7's on the table at the same time. If she can't pick up any cards from the table, she lays one card face up on the table.

Advanced: Besides playing according to the rules above, a player may build. For example, the player has a 2 and a 7 in her hand, and a 5 lies on the table. She may place the 2 on the 5, saying "I am building a 7." No player may pick up this 2 or this 5 alone. On the next round she picks up the 2 and the 5 with her 7, placing the 7 on top of her bundle.

Stealing the Bundle: Besides following the rules above, a player may capture another player's bundle by matching the top card in the bundle with a card in his or her hand. The player places the captured cards on top of his or her bundle.

Continue the Game: On each round a player removes exactly one card from his or her hand, by either matching it with other cards or laying it on the table.

After the first hand, deal cards only to the players, not to the table.

Winning: The winner is the player who has the largest bundle after all the cards have been played.

give change for a dollar in payment of a fifty-nine-cent item, I said aloud: "59, 60, 65, 75, one dollar," as I handed the coins to the customer. So when my fourth-grade teacher gave me two dollars and told me to "ask your father to choose a box of handkerchiefs, and be sure not to lose the change," I announced proudly that I could carry out the whole transaction myself.

What better way to acquire mathematical self-confidence! Not only did my parents trust me to handle merchandise and money, but I also was praised by the customers, who, as I look back, did not seem reluctant to be served by a child in elementary school. I even picked up some Ukrainian and German words in the process.

Although I was very comfortable with numbers and linear measurements, I had less aptitude for three-dimensional tasks. It might have been different if I had spoken up. When I was six, my mother gathered the stuff to be thrown away prior to our move to a new home. In the pile was a bag of long-forgotten alphabet blocks. I sat down in the midst of the rubbish and proceeded to build towers and other fanciful structures. But it never occurred to me to ask Mom to save this treasure.

Few people have been blessed with the kind of family encouragement that eventually led Patricia Kenschaft (age about 50) to her present position as professor of mathematics at Montclair State College in New Jersey. When she was four and a half, her father demonstrated with the dishes on the table that the ratio of the circumference to the diameter of a circle is always the same number, called *pi*. By second grade her mother had taught her the area of a triangle, the commutative property of multiplication (for example, 3×4 is the same as 4×3), and an algebra game, all such great fun! At the age of ten she expressed a desire to learn algebra. Her mother guided her to the algebra section of the encyclopedia, and she learned all of first-year algebra from that source. Her self-confidence and ability to study on her own enabled her, later in life, to earn a doctorate in pure mathematics while her two children were small.[2]

Ed (age 30) started college just recently. In high school he had struggled through elementary algebra—"I just could not make head or tail out of anything. My first year in college I started right at the bottom at the basic math level and was very surprised at how well I picked up on the concepts." Ed was encouraged by his stepfather, an engineer. "He has inspired me to the point that engineering is my main focus at this point in my life. I believe I would have pursued this avenue eventually, but it always helps to have someone to push you just a little further."

Joy (age 22) is an engineering major. She, too, has been motivated by her family. "My family is somewhat math oriented. My father excelled in math as a youth. My two uncles also excelled in math to the degree of obtaining their Bachelor of Science in different engineering disciplines." Joy believes that "a child's first experience with numbers should be a comfortable confrontation and not confusing. I believe that it should start in the home as a child and continue throughout elementary with the help of caring and capable teachers. This will enable one to have a more comfortable outlook on mathematics and may motivate one to pursue a math-related area."

Dr. Kenschaft and I both enjoyed extraordinary childhood opportunities to learn mathematics. Ed and Joy also gained inspiration and motivation from their families. But all parents and caregivers, no matter how slight their own math backgrounds, can encourage their children to enjoy mathematical activities and to do well in school math.

Importance of Family Involvement

Parents, naturally, want their children to achieve at the highest level. But many parents are up in the air when it comes to helping their children to do well in math. Perhaps they themselves are math fearers or math avoiders. They may think of math as merely dull, dreary drill of memorized facts and rote procedures, and feel incapable of presenting other kinds of math activities. Or they may believe that it takes a born genius to master a "hard" subject like math: "I couldn't do math, so I don't expect Andy to be good at it, either."

Eartha (age 19) was having no end of trouble with college algebra, following bad experiences with math in high school. "No one in my family is good with math so they didn't get mad if I got a bad grade. Didn't get much motivation [from my family]." I assume that Eartha's family expressed their negative feelings about math. Getting mad at her might not have helped, but friendly encouragement probably would have made a difference.

Family involvement in children's education is vital. Parents are in the best position to foster children's vast potential for learning, to inspire children's confidence in their ability to learn, and to develop favorable attitudes from the earliest age. It may seem far-fetched, but it is true that the way you treat your four-year-old may affect that child's SAT scores twelve years later.

A survey of thousands of students in Montgomery County, Maryland, indicated that one factor above all others motivated students to pursue mathematics in high school. These students "pointed to their parents as the primary force behind their interest in mathematics." They spoke of the encouragement that their parents provided, and the expectation that they would continue in math and do well in the subject.[3]

Parental encouragement is particularly important for girls. A survey commissioned by the American Association of University Women found that girls' self-esteem drops dramatically as they approach the adolescent years. White girls lost their self-assurance earlier than Hispanic girls, while black female high school students were the most self-confident. Boys fared much better than girls in this respect. Is it any wonder that girls lose their self-esteem in the face of the messages they receive from school authorities, the curriculum, and the society at large telling them that they are not worthy? Apparently black girls are better able to withstand the negative impact of these messages because of the strong support they get from their families.[4]

Reginald Clark's research with African-American families in three inner-city Chicago neighborhoods indicates that high-achieving children had parents who were involved in the children's home activities, studying and reading with them. Such factors as social class and single-parenthood were far less important in explaining achievement.[5]

Parents can boost their children's self-esteem and enable them to withstand the effects of teachers' low expectations. The earlier you take on the responsibility, the better your children will achieve in school. When children begin to fall behind, their deficiencies are magnified as they grow older, unless there is prompt intervention.

Herbert P. Ginsburg, professor of psychology and education at Teachers College (Columbia University), believes that "education is more than learning in the purely cognitive sense. Emotion is also part of the intellect." His studies show that the way a child approaches school learning is influenced by parents and affects the child's expectations in that area of learning. "Many Asian parents today are unschooled. . . . While many of these parents cannot help their children with academic material, they have a respect for learning and expect their children to do well in school." It is less the actual teaching and more the attitude and expectations of the teacher or parents that count.[6]

American students make a poor showing in mathematics in comparison with their Japanese counterparts. Often overlooked is the attitude of their parents. The assumption in Japan is that anyone who works at it can learn math. And Japanese children do work at it, sometimes to the point where, in the upper grades, the strain endangers their health. In the United States, on the other hand, the ability to do math is considered an inborn trait that some people have and others don't.[7] As we have seen in earlier chapters, there is little evidence for this assumption, and Japanese children offer a good counterexample.

Some parents may feel that they are not qualified to give their children a good start in math. Nothing can be further from the truth. Adults who have experienced difficulty in math are likely to be sympathetic to the problems their children face. All parents can give their children understanding and love, the first prerequisites for developing good attitudes. They can instill the self-confidence that will inspire a child to say: "I can do it!"

Nor is it necessary to follow the school curriculum. One of the best ways to involve children in math is to play games, all sorts of games, from Ring Around a Rosie to chess, depending on the child's age and interests. You don't need to know formal mathematics to play. Other family members and friends can join in, making it fun for all.

The home is the world's largest and best school system. Parents as teachers are not burdened with twenty or thirty students, nor are they regulated by school bells and administrative duties. Home is the ideal setting for children to learn attitudes and concepts, how to ask questions, how to seek and verify knowledge. Learning how to find answers is a skill that will serve for a lifetime.

Helping Preschoolers

Children are naturally curious about the world they live in. They love to learn new words, to play with new ideas, to investigate all sorts of situations. You as a parent can foster this vast potential for learning.

Babies pick up a great deal of math. One of their first words is "more," followed closely by the negative "no more." They can compare collections of objects, they know that one doll is bigger than another, and that one truck is heavier than another. They learn to recognize patterns, to repeat words in rhythm, and to carry out a series of actions in correct sequence. These are all mathematical ideas that parents can build upon.

Ordinary household tasks provide a setting for a variety of mathematical experiences. Sorting the laundry, matching socks by color, size, and pattern, setting the table, preparing food, shopping, fixing things—all afford opportunities for doing math while engaging in useful tasks. The job may take a little longer at first and there will be mishaps. But the outcomes in terms of children's experience and sense of accomplishment are worth all the time and effort.

We often think of counting as a child's first mathematical activity. There probably are more counting books than all other children's math books put together. Learning to count is a complex activity, and memorizing the number sequence "one, two, three, . . ." is only one step. In addition to knowing the words in the correct order, children must note each object once and only once, and assign one counting word to each object. They must recognize that the last counting word also tells how many objects have been counted, and that the order in which they count the objects can vary. Having children use their fingers and various materials as counting devices can be helpful. You should not be surprised by the child's inconsistent answers. It can take years for a child to learn all these concepts.[8]

Of all household activities, few can equal food preparation as a medium for learning math. Children practice counting, measuring liquid and dry substances, weighing, sorting, dividing, estimating, telling time, learning shapes, and recording.

Cooking is a useful skill for boys as well as for girls. And to avoid gender stereotyping, girls should have opportunities to play with trucks, build with large blocks, and conduct all sorts of experiments. This goes for the nursery school as well as the home.

Although you may not think of a hand-held calculator as a suitable toy for a preschooler, your child will probably enjoy pressing the keys to see what numbers will appear. Eventually, perhaps with a few hints, she will figure out combinations of numbers and operations that make sense to her.

You might ask your child to estimate how long a piece of string would just fit around her waist. Cut the string as she suggests and let her test her estimate. Then ask her to do the same with your waist. In fact, you might try it yourself! Repeat the procedure with other parts of the body and other objects.

This activity of "Guessing the Length" enables children to

Kids' Stuff

Four kids want to share a bag of apples. There are seven apples in the bag. How can they do it so that each kid gets the same amount?

Ancient Egyptian solution: The ancient Egyptians would have followed this procedure, as described in the Ahmes papyrus, written about 1650 B.C. by the scribe Ahmes (or Ahmose). First give each kid one whole apple. Then cut two of the remaining apples in half and give each kid a half. Cut the last apple into four equal parts and distribute them. Each child receives (1 + ½ + ¼) apples. Each child has three pieces of the same size and shape. The ancient Egyptians worked with unit fractions, fractions having the numerator *one.*

Modern kids' solution: Make applesauce and spoon out equal amounts to each kid.

learn that it's okay to risk making a wrong guess, that they themselves can check and correct their solutions to problems, and that they can become better estimators with practice. They learn that adults can also be off the mark in their estimates, that adults are not always the final authority.

Children actively construct their own knowledge by relating new information to what they already know. As they discuss their observations, they are learning to put their thoughts into words and to use mathematical language. Above all, they are developing the skills of analysis and critical thinking. Your role is not to instruct in the old-fashioned lecture method, but to encourage and respond to your child's efforts at learning.

If you can count in a foreign language, if you know songs and games from your own culture, you have an additional advantage. You can pass along to your children a valuable heritage in which they can take pride, and which they in turn can share with others.

I first encountered Carmen in the library of a local college. What caught my attention was that she was showing her four-

year-old daughter how to write answers in a first-grade math workbook so that she herself would be free to do her library research. The workbook task involved writing numerals, a skill that the child had not yet developed. I spoke up when I heard Carmen correcting her daughter because she had written the wrong numeral, although she knew the correct answer.

Carmen was doing the best she could. She told me that she was afraid of math, a fear that developed when as a child in Puerto Rico she lived in a "home" due to her mother's illness. "Now, when I see math in my chemistry course, I get scared." She spoke with pride about her daughter. "I don't want her to have these fears, and I'm trying to give her a good start. She learns so fast! I explained addition and subtraction, and she picked up the concepts in two minutes. Two minutes! But I really don't know how to help her."

Tyrone (age 21), an engineering major, wrote about his early introduction to math. "When I was about five years old my mother brought me a second-grade math workbook that was fun to use. However, I worked about halfway through the book and then got bored. . . . I really never had a fear of math. That early workbook solved that problem. My father was pretty good in math, but it was my mother who introduced me to that math workbook, probably because I was giving her a hard time. I will always thank her for it. So I feel the way to alleviate a math fear is to deal with it at an early age."

Helping School-Age Children

Children enter school with a wealth of mathematics knowledge. It is unfortunate that school authorities and teachers don't always recognize these achievements. They treat children as though they are empty containers into which knowledge must be poured bit by bit. Often the knowledge that children bring with them is simply ignored because it doesn't fit into the school curriculum. Although educational psychologists are well aware of children's strengths, changing the curriculum and some teachers' attitudes is a slow process.[9]

As children try to accommodate themselves to a rigid curriculum and the discipline that some teachers demand, they begin to lose their creativity. Drilling for the upcoming standardized test is hardly a stimulating activity. The need to use the one correct procedure to find the one correct answer has a stultifying effect on children's thinking processes.

This is where parents and caregivers come in. They can stimulate their children by playing games, posing problems, taking trips to museums and other places of interest. Reading and interpreting a road map, for example, is a great mathematical activity. It involves following a route, estimating, measuring and comparing distances, reading and interpreting scales, and much more. Map reading becomes meaningful when children can trace the route of an actual trip and read the road signs on the way. And don't leave out the girls. Research shows that girls are less exposed than boys to such activities. Many adults, men as well as women, drive cars, yet are unable to read road maps.

With stepped-up pressure on schools and teachers to improve test scores comes more emphasis on doing paper-and-pencil arithmetic before children are ready. Educators blame "highly competitive parents who, as their children enter nursery school or kindergarten, already worry about their future scores on the Scholastic Aptitude Test and admission to college. They want their children to be pushed academically."[10] Parents expect kindergartners to bring home stacks of papers to show that they have really learned something. Yet, children in more flexible schools achieve at least as well on standardized tests, and often better

Do they realize that the pennies would add up to over $10 million each month?

than the students in accountability-driven schools. Parents must understand how children learn and must relax the pressure.

What about homework? How much help should a parent give? Generally, help should be limited to encouragement and to seeing to it that the homework is done. Some schools give no homework in the early grades, and that may be just fine. If homework consists of a half-hour of dull, dreary, deadly drill on a single operation with numbers, it is time to raise objections with the teacher and the school. Children should be experiencing a variety of mathematical activities. Schools that offer less are depriving children of the opportunity to develop the necessary higher-order thinking skills.

Mai (age 26), a college student, has fond memories of her family's involvement in her math education in Vietnam. "Sometimes, if I have a problem that I can't find solution, I will discuss in my family to solve this problem. This discussion is very noisy and [they suggest] a lot of ways to find solution. My family think that math is necessary in our life and it is also a fun game in free time."

For Mark (age 19), math was not much fun in the early grades. To counter his lack of interest in the subject, "my mother practically drilled me to learn math." Although, as an engineering major, he regrets not having taken trigonometry in high school, he adds: "Today I'm glad my mother drilled mathematics in me because I would surely be lost without it. . . . I would like to say that mathematics will exist forever and we should learn all about it as much as we possibly can so we will be able to solve problems of tomorrow."

Some children need more help than their hard-pressed teachers can offer in class. One topic that causes difficulty in the early grades is place value, the key to understanding our system of numeration. Place value depends upon grouping by tens, just as ten pennies are equivalent to one dime, and ten dimes equal one dollar. Experiences in grouping and exchanging real pennies, dimes, and dollars can lead to activities with written numbers.

Children should also practice with other types of groupings. For example, how can nine coins be combined to make one dollar, using any of the following coins: pennies, nickels, dimes, quarters, and half-dollars? This is the kind of problem that has more than one correct answer; I found four. Children should experiment with actual coins before writing down any solutions. Questions like: "How many pennies can you use? Can you use exactly four pennies to combine with other coins? Why or why not?" will help them to analyze the problem and to make efficient choices. Putting their ideas into words encourages them to think clearly. You can assist your child in planning the trials in an orderly way and recording them in a table. Of course, children should write down all the trials, not just the successful ones! It is not a disgrace to try out a course of action and then find that it doesn't work. Trial and error is a legitimate method of problem solving. This is a fun way to practice operations with numbers and mental arithmetic. An introductory problem might be: "In how many ways can you combine pennies, nickels, and dimes to make a quarter of a dollar?" Children can invent their own problems, similar to these or different.

Pennies 1¢	Nickels 5¢	Dimes 10¢	Quarters 25¢	Half-dollars 50¢	Number of coins	Value in cents

Writing Two-Digit Numbers

What You Need: Ten dimes, ten pennies, pencil, paper

How to Play: Tell your child, "We are going to write how much money we have." Make a group of two dimes and four pennies. "How many dimes? . . . How many pennies? . . . I will write 2 in this column, and 4 in that column. This T stands for Ten; each dime is worth ten cents. This O stands for One; each penny is worth one cent.

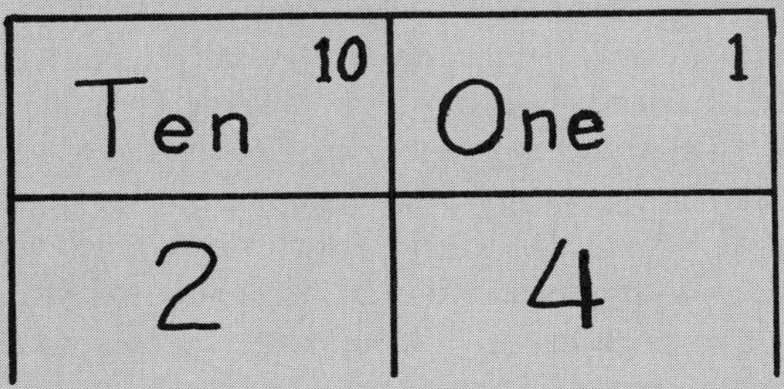

Ten 10	One 1
2	4

"Now tell me how many cents I have here. . . . Twenty-four cents. This number is twenty-four." Point to the written number. "It means two tens and four ones. Twenty-four."

Repeat with many other sums of dimes and pennies.

What Else Can You Do? Write a two-digit number. Ask your child to give you that much money in dimes and pennies. Take turns.

(From Claudia Zaslavsky, *Preparing Young Children for Math: A Book of Games* [New York: Schocken Books, 1986], 107.)

Parents, Watch Your Attitude

It is difficult for parents to see their children make mistakes and not jump in to correct them immediately. It's wise to exercise restraint. Children who are watched as they work and who are criticized for their errors may develop a fear of taking risks, of exploring new possibilities. They may even begin to hate math. A more fruitful course would be for parents to ask questions and give hints that will lead children to discover their own errors. Adults must respect children's ability to take responsibility for their own learning.

Janet (age about 40), a nursing student, was enrolled in an algebra class. Although she knew the material well enough to help other students in the class, she received only 35 percent correct on the first test. When the instructor tried to determine the reasons for her poor performance, Janet became so distraught that she barely knew the sum of two plus two. Finally she blurted out: "My father has been dead for ten years, but he is standing behind me criticizing everything I do."[11]

The early home experiences of Nita and Lola offer a direct contrast. Both are successful women, about forty years of age at the time they wrote their autobiographies. Both were studying or working in math-related fields. Nita was studying for a master's degree in architecture. Lola directed the math/science tutoring lab at an inner-city college.

Lola speaks fondly of her parents' attitude:

> When I try to understand why I like math, it is obviously because I never bought the school's definition of what the subject was all about. My home experiences with math were always pleasant. I remember feeling deep satisfaction at being able to master the times tables under my mother's guidance. I would go to my bedroom, study a table, and then report to the kitchen, where she was cooking, for a recitation. Errors were handled in a relaxed manner, and suggested strategies for studying were appropriate for my level. I never remember any exasperation on her part, only patience (I guess I was a lucky kid). My father had an equally positive attitude towards math. Although he only finished tenth grade, he frequently told me about his pleasure with algebra. He particularly liked working out the equations. . . . My experiences throughout high school were boring. But I continued to do well, for how can you go wrong when all you are asked to do is memorize the rules?

Learn from Mistakes

A busy teacher might check the answers to the subtraction exercises below and mark them wrong. All the answers are wrong, but for just one reason. They show a common error based on failure to understand place value and "regrouping"—also called "borrowing." In each column the smaller digit was subtracted from the larger digit.

16	24	230	732
− 8	−15	− 13	−359
12	11	223	427

One way to remedy the situation is to have your child practice with concrete materials. Assemble a collection of dimes and pennies. On a sheet of paper draw two columns, headed Ten and One, as on page 207. Have your child carry out a subtraction problem that requires regrouping, first with the coins and then on paper. Exchanging a dime for ten pennies is like regrouping in a pencil-and-paper exercise. Repeat with other numbers. Pretend that you are making purchases at a store, to give it a real context.

Once your child has caught on to subtraction of two-digit numbers, go on to three-digit numbers, using dollar bills, dimes, and pennies. On paper, head the three columns Hundred, Ten, and One. Take turns making up problems and solving them.

Nita holds her father responsible for some of her difficulties with mathematics: "My father was very demanding and easily exasperated by my poor performance in arithmetic. Algebra was even worse. You can imagine how he carried on when I failed algebra four times in high school! I couldn't learn it at all. It was all memorization, and I can't memorize. I love geometry—it's concrete, I can see it—but because I hadn't passed algebra I wasn't allowed to take geometry until my senior year. My grade was 87 percent in the course. Now I am finally studying what I like." A deeper analysis might reveal to what extent her father's overcritical reactions influenced her ability to learn algebra. Certainly he was not helpful. Eventually Nita did master the algebra required for her degree in architecture, but at what great cost!

FAMILY MATH to the Rescue

FAMILY MATH is a wonderful new development in mathematics education. FAMILY MATH means parents and their children doing mathematics together and having fun doing it. FAMILY MATH is about helping people to like math and to succeed at it. FAMILY MATH is the program that enabled Mary Jo Cittadino to overcome her fear and avoidance of mathematics as she organized classes for other parents and their children.

The object of the program is to create a friendly environment where parents and other caregivers work together with children on various activities designed to include and supplement the school math curriculum. The activities involve doing, rather than memorizing, and enable children to see how math and daily life are intertwined. They might survey the group to find out each person's favorite ice cream flavor, then make graphs to illustrate their findings. They talk about symmetry while making decorative designs. They analyze the strategies involved in the games they play. They discuss puzzles requiring logical solutions. They even create their own puzzles and games. While having fun in a cooperative atmosphere, they deal with many mathematical topics—arithmetic, geometry, measurement, probability, statistics, graphing, calculators and computers, and logical thinking.

The activities require only inexpensive, easily available materials that the participants can take home for further practice. A FAMILY MATH book is packed with suggested activities for different age groups, from kindergarten through early high school grades.

Who are the leaders? They might be classroom teachers, par-

ents, senior citizens, church members—in short, anyone can become a leader. They may choose to attend training sessions, or just use the FAMILY MATH curriculum guide as a basis for their activities. The guide is available in Spanish and other languages, as well as English. A seventeen-minute film, *We All Count in Family Math,* is a further aid to participants. Classes may meet after school or in the evening, in various places throughout the community at the convenience of the participants.

FAMILY MATH motivates kids to want to take more math in school. They see that math can be both useful and fun. Women and men who work in fields requiring math and science are invited to speak about their jobs and to encourage children to think about math-related careers.

As one parent remarked, "It's too late for some people to go back; well, it's not too late for my children; that's why we have to try."[12]

What Else Can Parents Do?

Parents can help their children learn about careers and their requirements. Twenty years ago few people could foresee the technological breakthroughs and social trends that have led to whole new areas of employment. Who knows what the future will bring? Young people should keep their options open by taking as much mathematics as possible in school. Their test scores will improve and they will be prepared for a wide variety of college majors, leading to the careers of the future. See the "Resources" section for lists of organizations that can furnish such information.

The media rarely feature mathematicians and scientists. Rarer still on the news pages are mathematicians and scientists who are female or members of "minority" groups. Young people should learn that folks who look like them can be successful in these professions.

Parents should get to know their children's teachers, observe classes, volunteer to help, and become active in the parent-teacher organizations. Some teachers and schools discourage parent participation, and that is a real problem! What are they trying to hide? Parents can bring a different perspective, they can tell about their own experiences, they can share their specialized knowledge. Schools should provide workshops to inform parents about the math curriculum and suggest how they can work with their children. Of course, FAMILY MATH sessions would be ideal.

Parents need to keep a watchful eye on the schools and the kind of math that children are expected to learn. Is the math curriculum designed to encourage high-level thinking and love of mathematics, or do standardized tests dictate the curriculum content? Are students tracked into low-level courses and drilled in "basic skills"? Are the teachers qualified to teach the curriculum? What kind of homework is required? Are students receiv-

National Council of Teachers of Mathematics (NCTM) Policy Statement

As a professional organization and as individuals within that organization, the Board of Directors sees the comprehensive mathematics education of every child as its most compelling goal.

By "every child" we mean specifically—

- students who have been denied access in any way to educational opportunities as well as those who have not;
- students who are African American, Hispanic, American Indian, and other minorities as well as those who are considered to be part of the majority;
- students who are female as well as those who are male; and
- students who have not been successful in school and in mathematics as well as those who have been successful.

The Board of Directors commits the organization and every group effort within the organization to this goal. (NCTM *News Bulletin* [November 1990]: 3.)

The NCTM Policy Statement may be a useful tool in persuading your child's school or teacher to improve the teaching of mathematics and in working with the parent-teacher organization toward these goals.

ing adequate guidance as to courses, careers, and postsecondary education? For meaningful change to occur, parents need to work through their parents' associations and community group coalitions to monitor achievement levels and school practices, and to become involved in the political process.

Some of the new programs that I have described have a specific parent component. The directors of The Algebra Project emphasize that parents must be involved if efforts to improve

SMART Girls

Operation SMART is a successful out-of-school program to encourage girls in Science, Mathematics, and Relevant Technology (SMART). Established in 1985 by Girls Inc. (formerly Girls Clubs of America), the program offers hands-on experiences that involve girls in the sciences and stimulate them to continue their study of science and math in school.

Evelyn Roman-Lazen, the Operation SMART coordinator, exemplifies in her own person the positive influence of the program. Born in the United States, she spent much of her childhood in Puerto Rico, returning to the mainland in 1988 at the age of twenty-two. Poor teaching in elementary and junior high school caused her to hate math. She admits that she cannot remember the multiplication tables. Although she chose business courses in high school, her older siblings advised her to take algebra and geometry so that she might enter college.

Much to her surprise, her major in marketing and economics at the University of Puerto Rico included many required courses in mathematics—calculus, differential equations, linear algebra, mathematical statistics, econometrics. Later, in her job at a large hotel, she used her mathematical knowledge to work out strategies for the operation of the casino. She was not happy in this job—"figuring out how to empty people's pockets." Besides, she was a victim of sex discrimination in salary.

Her first job in New York was as an administrator with Girls Inc. Contact with Operation SMART inspired her to continue her studies in math and to take courses toward a degree in electrical engineering. Now she trains leaders and teachers to work with Operation SMART and Family Science programs. Her passion for math and science is unbounded, and her enthusiasm is an inspiration to others. (Personal communication with Evelyn Roman-Lazen.)

the mathematics education of children are to succeed. "Teaching algebra to parents empowers them not only to grasp mathematical concepts that may have been unfamiliar to them and reduce any associated mathematics anxiety but it also equips them to be active partners in their children's mathematics education."[13]

Young people who have out-of-school math experiences become enthusiastic about the subject and achieve better in school. These intervention programs are generally successful in engaging even the most turned-off students. Such programs are offered by universities, museums, schools, and various education and community organizations. They feature one or more of the following:

- Conferences on careers and colleges
- Field trips to science museums, zoos, and industrial sites
- Competitions and contests
- Academic instruction after school, on Saturdays, in summer, or in university courses
- Internships in industry and universities

See earlier chapters and the Resources appendix for information about such programs.

Parental involvement in these programs is encouraged and even required. Xavier University, a historically black institution in New Orleans, runs summer enrichment programs for secondary-level students, as described in chapter 5. The day before the program begins parents are expected to attend an orientation session. They receive report cards frequently and attend an awards assembly at the end of the program. All students receive certificates and have the opportunity to demonstrate what they have learned.[14]

The 1989 Report to the Nation, *Everybody Counts*, tells us: "Children *can* succeed in mathematics. Many do so in other countries and some do so in this country. The evidence from other nations shows overwhelmingly that if more is expected in mathematics education, more will be achieved. Clear expectations of success by parents, by schools, and by society can promote success by students."[15]

CHAPTER NINE

Mathematics of the People, by the People, for the People

> There is a sense that they are dealing with an issue that does not feel very vivid, and that nothing that we say about it to each other really matters since it's "just a theoretical discussion." . . . Questions of unfairness feel more like a geometric problem than a matter of humanity or conscience.
>
> —*Jonathan Kozol,* Savage Inequalities

In this excerpt[1] Kozol describes a discussion with students at a well-funded school in a wealthy New York suburb. The students are arguing that giving equal funding to schools in poor districts would not improve education in those schools. They present one argument after another to prove their points, just as though they were going through a logical proof in a geometry class. Notice how Kozol contrasts "a geometric problem" with "a matter of humanity or conscience." He is not referring to the geometry of the Palestinian woman fashioning a shirt without the aid of a pattern. He doesn't have in mind the geometry of an African family building their own home. With his analogy to geometry he wants to evoke the image of an attitude that is cold, abstract, unrelated to human concerns. But he doesn't need to explain his analogy. He can take for granted that his readers will know exactly what he means—that they, too, have experienced school geometry as a subject devoid of humanity.

We have discussed some of the reasons that mathematics is considered difficult and unattainable for all but the elite, mainly white middle- and upper-class males. One factor is the manner in which mathematics is taught. The Brazilian educator Paulo Freire talks about two opposing types of education: dominating and dehumanizing, or liberating and humanistic. In the first type, the teacher as the authority pours knowledge into the students, the empty vessel. Freire calls this the "banking" method of education. Information is "banked," or, as a student in Jean Anyon's study commented, it is put in "cold storage" for use at some future time. The experiences of the dominant culture are the only ones worthy of recognition, while those of other people are ignored or discredited. They are presumed to have no history or culture.

In contrast, humanistic education starts with the experiences of the learner. Teacher and student have equal status in this system. Together they engage in problem-solving activities, which they subject to critical analysis at every step. As the learners become aware of the forces in society that shape their lives, they begin to see that they themselves must become actively involved in the democratic process.[2]

Real mathematics education should fulfill the conditions of humanistic and liberating education. It should equip people with the knowledge and tools that will enable them to examine and criticize the economic, political, and social realities of their lives. It should help them to understand and try to find solutions to some of the hard questions that affect them, such as:

- Why are the taxpayers and their children and grandchildren stuck with paying a half trillion dollars for the savings-and-loan fiasco?
- Why does so little of the federal budget go to education, health care, and housing, and so much to building weapons and making the rich richer?
- Why is the infant mortality rate in the United States twice as high for black babies as for white babies?
- How can we deal with long-term unemployment? With homelessness?
- How can we clean up our polluted environment and stop further pollution?
- How can we save the planet and its people?

Many people are critical of mathematics, science, and technology as they are now practiced. They claim that an elite of predominantly white upper-class males has set the priorities, with little regard for the needs of human beings and the preservation of our planet. They deplore the fact that the most highly

Doublespeak

To hide the true state of the economy, government officials and business executives are finding new ways to use the vocabulary of mathematics. When a company demoted its employees, they called the process "negative advancement." The increasing trade deficit is an example of "negative growth." A newspaper article tells of a bank in the Sunbelt with a "negative net worth of $447.8 million." Later in the article we read: "The savings and loan industry's mounting troubles in Texas can be measured by its total capital, which stood at negative $10.1 billion [due to] bad loans, an overbuilt real estate market, fraud and mismanagement."*

Will the reader overlook the word "negative" and think that there is no cause for concern? What other reason can there be for this "doublespeak"? Of course, since the publication of that article in 1988, people have become aware of the full scale of the savings and loan debacle—or as much of it as they are allowed to find out.

*Thomas C. Hayes, "$1.2 Billion Saving Unit Loss Seen," *New York Times*, 14 May 1988, 37.

trained minds are working with the most up-to-date equipment to devise ever more destructive weapons systems, while billions of people are living in poverty and deprivation. In the late 1980s the United States invested more research money, both public and private, in military development than in the development of our resources for civilian purposes.[3]

James S. Jackson, University of Michigan psychology professor, believes that opening the university to women and to racial and ethnic minority group members can bring some badly needed fresh perspectives to the scene. In an article entitled "From Segregation to Diversity: Black Perceptions of Racial Progress," he wrote: "Blacks, other minorities and women bring different ways of representing and conceptualizing problems and of addressing intellectual issues. . . . [Their] experience can

The Rich Get Still Richer

According to an article in *Earthwatch* (July/August 1992) the rich are getting richer and the poor are getting poorer, not only in the United States but also in the world as a whole. In 1990 the income of the wealthiest one-fifth of the world population was about sixty times the income of the poorest fifth, compared with thirty times in 1960. If this trend continues, what will be left for the next generation of poor people in the world?

have profound effects on the way in which problems are conceptualized and how solutions are attempted. Just as we recognize that artists view the world differently, bringing us unique insights, I believe it is also true that in the humanities, social and physical sciences that different perspectives are needed."[4]

I must agree with Professor Jackson when, for example, I compare the present (1992) administration's national budget with the alternative budget proposed by the Congressional Black Caucus or the legislation supported by women in Congress. The African-American and female legislators, for the most part, support a reduction in weapons spending and an increase in funds for programs that will benefit society, rather than destroy it.

In an impassioned speech to an international audience in 1984, Australian mathematics educator Nancy Shelley condemned the culture of militarism, while calling for more humanistic applications of mathematics: "This culture of militarism is exercised, for the most part, by white men. . . . Today women, and people from other cultural backgrounds, are excluded [from decision-making] by the bogus construct of 'ability.' . . . Other systems of mathematics are possible, which offer the possibility of a different world. Women have the opportunity to develop a mathematics which can provide the world with another way so that the planet lives and people the world over are fed and healthy."[5] This is clearly a call to women and people of color—and ethical white men—to take the initiative and change the priorities of society to benefit people, not profits and war. Mathematics can play a crucial role in this conversion.

Making Sense of the Dollars

President Reagan stated in his television address on 22 November 1982: "In recent years, about one-quarter of our budget has gone to defense, while the share for social programs has nearly doubled."

One year later, Andrew Stein, president of the Borough of Manhattan in New York City, wrote in an OpEd in the *New York Times* (20 November 1983): "Last year . . . defense expenditure soaked up 59 percent of the nation's general revenues, while social programs accounted for 18 percent."

Who was telling the truth? Perhaps both. It all depends upon how you look at the question.

Let's take a look at the federal budget for 1985. I have worked extensively with the figures for that year, and they are not much different from those of 1982—or of 1992, for that matter.

The total administration request for budget authority—that's what the president asks Congress to spend—was $925 billion in 1985. The first pie chart shows how

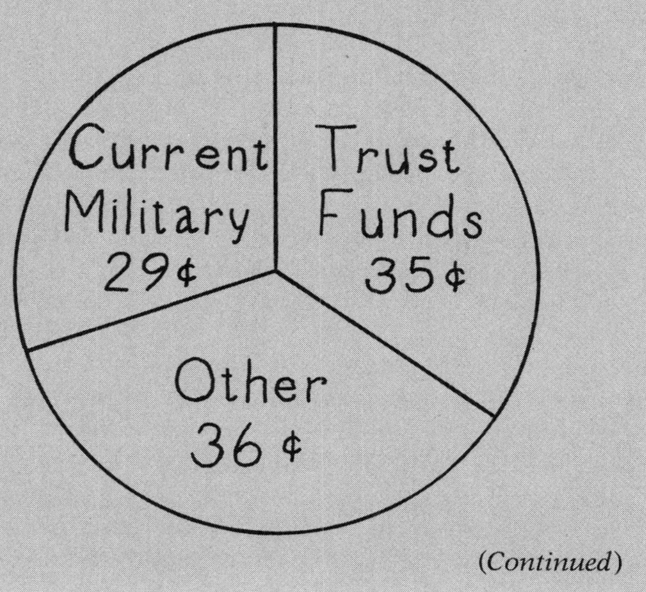

(*Continued*)

Making Sense of the Dollars (*Continued*)

this sum was split up. For every dollar requested, 29¢ was for current military, 35¢ for trust funds, and 36¢ for other purposes.

But trust funds are mainly for social security and are held in trust for people. Congress is not supposed to spend this money in other ways. Before 1969 these funds were not included in the federal budget. Putting them into the budget makes the military portion seem smaller.

Let's remove trust funds from the budget. The next pie chart shows how the federal budget dollar is apportioned when we exclude trust funds: 44¢ for current military, 22¢ for interest payments, 23¢ for human resources, and 11¢ for other purposes.

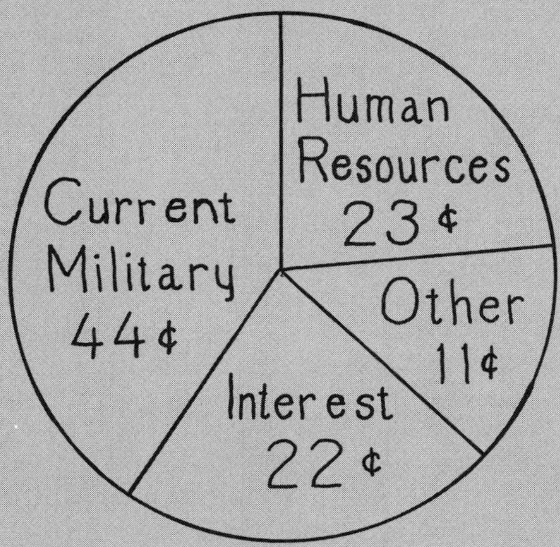

But there is more to the story. A large portion of the interest the government pays out on the national debt is due to borrowing for military purposes in the past. Veterans' benefits and services are also military. Now we have a third breakdown, as shown in another pie chart. Overall

(*Continued*)

Making Sense of the Dollars (*Continued*)

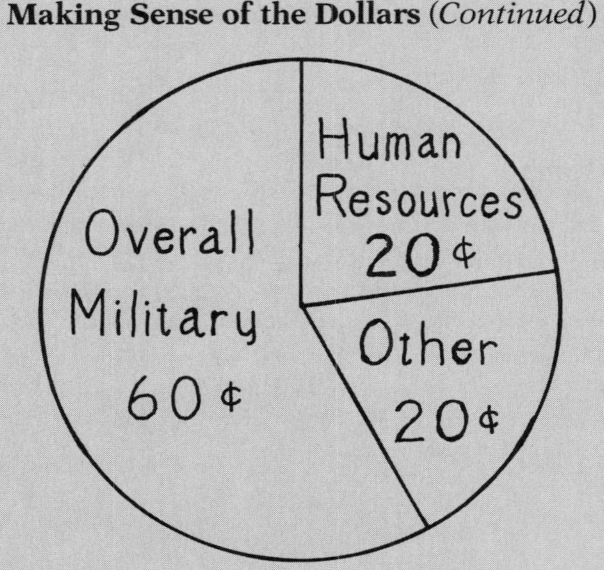

military is now 60¢ out of the federal budget dollar, while 20¢ goes to human resources and another 20¢ for other purposes.

So who was right, President Reagan or Andrew Stein? According to the official budget figures for 1982, just about 25 percent went to the category called "Current Military." President Reagan was right on that score when he said: "About one-quarter of our budget has gone to defense." However, when we analyze the figures again after removing trust funds and including all military expenditures, we see that Stein was right when he said: "Defense expenditures soaked up 59 percent of the nation's general revenues, while social programs accounted for 18 percent."

In the early 1980s Nancy Shelley gave up her full-time work in mathematics education to devote herself to working, writing, and taking action for peace. The 1991 war in the Persian Gulf horrified her, as it did many others, with its dehumanization of the people under attack and "humanization" of the attack weapons, as she commented in a 1992 speech entitled "Mathematics: Beyond Good and Evil?":

> While the Iraqi people faced the reality of being bombarded, on TV screens throughout the world, viewers were bombarded with a virtual reality display of the "theatre" of war. Weapons took on human attributes: "smart" weapons had "eyes," computers "brains" which "made decisions," and it was military targets and missiles which were "killed." Human beings acquired the characteristics of objects: if they were military personnel they "absorbed munitions"; if civilian, they were not hurt, maimed or killed but recorded as part of "collateral damage."[6]

In this speech Shelley reaffirmed "the possibility of a different mathematics which grows out of a culture which has no pretensions to dominance; which can yet engage a learner creatively; which affirms humanness; and which attends human needs and will embrace earthy things. A culture of relationship, of care, of communion. The process by which it will grow is known, the shape it will take remains to be made."[7]

Many scientists and mathematicians do feel responsible for the ultimate uses of their work, as exemplified by the significant protest against the Strategic Defense Initiative (Star Wars) in the 1980s, when thousands of scientists and mathematicians, among them Nobel Prize winners—and they are mainly white males—registered their opposition to this trillion-dollar program of destruction and vowed not to participate in it.[8]

The priorities of our policymaking bodies can and must be changed, if we are to survive. People have the power to make these changes. Just as slavery was abolished in the United States after years of struggle, so science, mathematics, and technology must be turned around to fulfill their potential of serving all of humanity.

I worry, and I am angry, thinking about the future of our children and grandchildren. I am sure that you share that concern. If this book has given you insights on what to do about the things that matter, then I will feel it is a success.

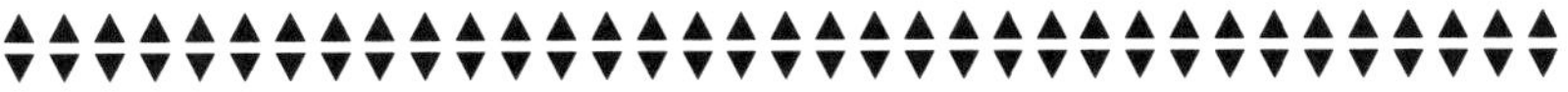

Resources

Intervention Programs

Intervention programs—math programs that take place outside of the regular school setting—are generally successful with female and minority students, even with those who do poorly with the school mathematics curriculum. What factors account for the difference?

Intervention programs include several features that are often missing from school math, among them:

- Well-trained teachers with high expectations for all students
- Cooperative-group work on relevant projects that motivate students to learn the math they need to solve problems as they arise
- Parental involvement
- Career information and role models
- Encouragement to continue the study of mathematics

The logical conclusion is that school programs should encompass all these features. This would entail greater funding, better teacher education, school restructuring, different forms of evaluation, and a revised attitude toward mathematics and mathematics education on the part of all sectors of the population.

Precollege Programs

It would be impossible to list all, or even many, of the current precollege intervention programs. Many programs are sponsored and/or funded by the organizations and government agencies listed in the next section. Four useful guides are the following:

Clewell, B. C., Bernice T. Anderson, and Margaret E. Thorpe. *Breaking the Barriers: Helping Female and Minority Students Succeed in Mathematics and Science.* San Francisco: Jossey-Bass, Inc., 1992.

Clewell, Beatriz C., Margaret E. Thorpe, and Bernice T. Anderson. *Intervention Programs in Math, Science, and Computer Science for Minority and Female Students in Grades Four through Eight.* Princeton, N.J.: Educational Testing Service, 1987.

Kenschaft, Patricia C., and Sandra Z. Keith, eds. *Winning Women into Mathematics.* Washington, D.C.: Mathematical Association of

America, 1991. Includes brief descriptions and addresses of typical programs at various grade levels.

Educating Tomorrow's Engineers: A Guide to Precollege Minority Engineering Programs. New York: National Action Council for Minorities in Engineering (updated frequently). Directory of over one hundred educational initiatives, organized by states, designed to build the pool of African-American, Hispanic, and American Indian students to enter technical fields.

Several programs have a national scope, among them:

- Expanding Your Horizons in Math and Science career conferences for girls and young women. (See Math/Science Network)
- Family Math, described in chapter 8. (See EQUALS)
- Girl Scouts of the USA encourages math-related activities.
- Operation SMART (Science, Math, and Relevant Technology) and SMART Eureka aim to encourage girls of all ages to continue in math and science. Local clubs develop participatory programs that enable girls to carry out interesting projects and explore career possibilities. (See Girls, Inc.)
- Parent Teacher Associations distribute mathematics kits containing ideas and materials for parent/child activities.
- Saturday academies, after-school programs, and summer programs. Contact local schools and colleges.

College Programs

Programs with a national scope include the following:

- Comprehensive Regional Centers for Minorities. Funded by the National Science Foundation (NSF), to promote science education for students from elementary school through college, train science teachers, and foster community involvement. Contact NSF, Undergraduate Science, Engineering, and Mathematics Education, ACCESS programs.
- Math Without Fear, Math Anxiety, Mind Over Math. These clinics hold sessions for adults to help them become comfortable with mathematics. Some programs concentrate on the psychological aspects, some on math tutoring, while others combine both aspects. Contact local colleges.
- Minority Engineering Programs on many college campuses. Contact NACME.
- Professional Development Program, described in chapter 7. (See Dana Center for Mathematics and Science Education.)
- Summer bridge programs for entering students. Contact local colleges.

RESOURCES

Organizations and Government Agencies

The following organizations fund and/or sponsor programs designed to foster interest and success in mathematics and math-related careers among women and minorities. Contact them for information about their publications and resources.

Academies of Science (local and state) sponsor programs.

Algebra Project (see chapter 6), 22 Wheatland Avenue, Boston, MA 02124; (617)287-1508.

American Association for the Advancement of Science (AAAS), Directorate for Education and Human Resources, 1333 H Street, NW, Washington, D.C. 20005-4792; (202)326-6670. Major objective: "To affect the quality of science, mathematics, and technology education in formal and informal education . . . for all people." Initiates projects, organizes conferences, and publishes reports, books, and newsletters directed to various segments of the population, including parents, with the goal of promoting the equitable representation of all people in mathematics and science.

American College Testing program (ACT), P.O. Box 168, Iowa City, IA 52243. Information on college entrance and other tests.

American Indian Science and Engineering Society, 1085 Fourteenth Street, Suite 1506, Boulder, CO 80302; (303)492-8658. Encourages participation of American Indians in technical fields, starting with early secondary grades.

Association for Women in Mathematics (AWM), 4114 Computer and Space Sciences Bldg., University of Maryland, College Park, MD 20742-2461; (301)405-7892. Supports women mathematicians, publishes newsletter and a variety of resource lists. Speakers' bureau serves high schools and colleges.

College Board, 45 Columbus Avenue, New York, NY 10023-6992; (212) 713-8000. Information about the Scholastic Aptitude Tests and other tests; intervention projects and inspirational materials; EQUITY 2000 project aimed at preparing minority students for college by taking academic high school mathematics courses.

Consortium for Educational Equity, Rutgers University, 4090 Livingston Campus, New Brunswick, NJ 08903; (201)932-2071. Gender Equity Unit of the Equity Assistance Center for Region B (see Desegregation Assistance Centers, below). Excellent resources: lists of materials and books, lending library, conferences and programs.

Dana Center for Mathematics and Science Education, 230B Stephens Hall, University of California, Berkeley, CA 94720. Sponsors the Professional Development Program described in chapter 7.

Desegregation Assistance Centers. Ten federal centers, each serving a different geographic region, provide services in race, gender, and national origin equity, conduct programs, and publish a variety of materials. For addresses and further information, contact the Office of

Elementary and Secondary Education, Department of Education, 400 Maryland Avenue, SW, Washington, DC 20202-6330; (202)401-0358.

Education departments of each state and city. Most maintain Offices of Equal Opportunity.

Efficacy Institute, Jeffrey P. Howard, President, 99 Hayden Avenue, Lexington, MA 02173; (617)862-4390. Works with school systems and organizations to "advance the intellectual development and academic performance of all students, with a particular focus on the development of African-American and Hispanic students." (See chapter 4.)

EQUALS, Lawrence Hall of Science, University of California, Berkeley, CA 94720, (510)642-1823. Conducts programs at many sites around the United States and elsewhere, for school personnel, students, and parents, to promote the participation of female and minority students in mathematics and computer education. Publishes print and audiovisual curriculum materials and career information, such as *Family Math* and *Math for Girls and Other Problem Solvers.*

Girl Scouts of the USA, 420 Fifth Avenue, New York, NY 10018; (212)852-8000. Publishes resource materials and sponsors events to encourage girls in math, science, and technology; conducts activities leading to math badges.

Girls, Inc., 30 East 33rd Street, New York, NY 10016; (212)689-3700. (See Operation SMART, under Precollege Programs.)

Higher Education for Working Adults, College and University Department, 555 New Jersey Avenue, NW, Washington, DC 20001; (202)879-4422. (See chapter 5 for a discussion of colleges for working adults.)

International Study Group on Ethnomathematics. Studies the influence of sociocultural factors on the teaching and learning of mathematics; publishes newsletter. Contact the president, Dr. Gloria Gilmer, Math-Tech, Inc., 9155 North 70th Street, Milwaukee, WI 53223; (414)355-5191.

Math/Science Network, Preservation Park, 678 Thirteenth Street, Suite 100, Oakland, CA 94612; (510)893-MATH. Provides support and resources for programs to serve female students of all ages, parents, and educators, women reentering the academic world, and professional scientists; publishes newsletter; sponsors "Expanding Your Horizons in Math and Science" career conferences in many parts of the country.

Mathematical Association of America (MAA), 1529 Eighteenth Street, N.W., Washington, DC 20036; (202)387-5200. Maintains committees to improve the participation of women and underrepresented minority groups; supports "Women and Mathematics" (WAM) visiting lecture program and SUMMA (Strengthening Underrepresented Minority Mathematics Achievement), a many-faceted program; publishes *Winning Women into Mathematics,* the SUMMA *Resource Directory of Intervention Projects,* and other relevant literature.

Mathematical Sciences Education Board, National Research Council, 2101 Constitution Avenue, Washington, DC 20418; (202)334-3294. Assumes a national leadership role in the reform of mathematics education; publications include the very readable *Everybody Counts: A Report to the Nation on the Future of Mathematics Education* (1989) (order from National Academy Press at the above address).

National Action Council for Minorities in Engineering, Inc. (NACME), 3 West 35th Street, New York, NY 10001-2281; (212)279-2626. Dedicated to bringing the talents of African Americans, Hispanics, and American Indians to the nation's engineering workforce. Conducts research, provides scholarships and grants, develops and operates model intervention programs, and publishes original educational materials for educators and for students of all ages.

National Council of Teachers of Mathematics (NCTM), 1906 Association Drive, Reston, VA 22091; (703)620-9840. Publishes materials relating to precollege mathematics education—books, audiovisual materials, journals for several levels of mathematics education, and a research journal. The NCTM was the first professional organization to establish national standards for precollege education, with the publication of *Curriculum and Evaluation Standards for School Mathematics* (1989), described in chapter 6.

National Science Foundation (NSF), 1800 G Street, NW, Washington, DC 20550. Federally funded foundation initiates and supports research and education programs in science, mathematics, and technology at all levels. For general information, request *Grants for Research and Education in Science and Engineering* from the Forms & Publications Unit at the above address.

QEM (Quality Education for Minorities) Network, 1818 N Street, NW, Suite 350, Washington, DC 20036; (202)659-1818. Dedicated to improving education for minorities throughout the nation; focus on encouraging the greater participation of minorities in mathematics, science, and technology at all levels; information about exemplary programs. Send for comprehensive national plan: *Together We Can Make It Work: A National Agenda to Provide Quality Education for Minorities in Mathematics, Science, and Engineering* (1992).

Society of Hispanic Professional Engineers, 670 Monterey Pass Road, Monterey Park, CA 91756. Works to improve the quality of education and training programs for Hispanic students entering engineering and science fields.

United States Department of Education, Washington, DC 20202. Funds programs on all educational levels. Send for *Guide to Department of Education Programs.*

Women in Mathematics Education (WME), c/o SummerMath, 302 Shattuck Hall, Mt. Holyoke College, South Hadley, MA 01075; (413) 538-2608. Promotes mathematical education of women and girls; publishes newsletter and bibliography.

Women's Action Alliance, 370 Lexington Avenue, Suite 603, New York, NY 10017; (212)532-8330. Sponsors the Computer Equity Expert Project, to increase girls' involvement with computers, math, and science; holds conferences and publishes literature for educators and parents on girls and computers.

Women's Educational Equity Act (WEEA) Publishing Center, Education Development Center, 55 Chapel Street, Suite 271, Newton, MA 02160; (800)225-3088. Promotes educational equity for women by publishing a variety of curriculum, career guidance, and teachers education materials for all age levels, including *Math, Science, and Your Daughter* and three other pamphlets in the series, *Encouraging Girls in Math and Science,* by Patricia B. Campbell.

Distributors of Mathematics Books and Materials

Request catalogs, specifying your interest.

Activity Resources Company, P.O. Box 4875, Hayward, CA 94549; (510)782-1300.

Creative Publications (Order Department), 5040 West 111th Street, Oak Lawn, IL 60453; (800)624-0822.

Cuisenaire Company of America, P.O. Box 5026, White Plains, NY 10602-5026; (800)237-3142.

Dale Seymour Publications, P.O. Box 10888, Palo Alto, CA 94303-0878; (800)872-1100.

EXA Company, 6928 East Sunnyvale Road, Paradise Valley, AZ 85253; (602)948-6795. Distributor of Algebra Blox, a concrete approach to learning algebra, suitable for all ages (see chapter 7).

J. Weston Walch, 321 Valley Street, P.O. Box 658, Portland, ME 04104-0658; (800)341-6094.

Select Bibliography

The publications listed below comprise just a small fraction of the available offerings in each category. For the most part, I have chosen recent publications. You can find additional titles in the reference notes.

Equity Issues and Overcoming Fear of Math

Afflack, Ruth. *Beyond Equals: To Encourage the Participation of Women in Mathematics.* Math/Science Network, 1982. (Order from EQUALS.) Designed for use by instructors of adult women, the materials stress problem solving, concrete representations, logical reasoning, and spatial experiences.

Arem, Cynthia. *Conquering Math Anxiety: A Self-Help Workbook.* Pacific Grove, Calif.: Brooks/Cole, 1992.

RESOURCES

Brecher, Deborah. *The Women's Computer Literacy Handbook.* New York: New American Library, 1986. Clear explanations of computer operations, as well as aspects of computers that are of special interest to women.

Chipman, Susan, Lorelie Brush, and Donna Wilson, eds. *Women and Mathematics: Balancing the Equation.* Hillsdale, N.J.: Lawrence Erlbaum, 1985. Research into the many aspects of women's underrepresentation in mathematics courses and mathematics-related fields.

Clewell, Beatriz C., Bernice T. Anderson, and Margaret E. Thorpe. *Breaking the Barriers: Helping Female and Minority Students Succeed in Mathematics and Science.* San Francisco: Jossey-Bass, 1992. Identifying and breaking the barriers to achievement.

Frankenstein, Marilyn. *Relearning Mathematics: A Different Third R—Radical Maths.* London: Free Association Books, 1989. Societal factors that induce math fear, and ways that adults can learn and enjoy math as they deal with real-life problems.

Hilton, Peter, and Jean Pedersen. *Fear No More: An Adult Approach to Mathematics.* Menlo Park, Calif.: Addison-Wesley, 1983. The concepts that underlie arithmetic.

Kenschaft, Patricia C., and Sandra Z. Keith, eds. *Winning Women into Mathematics.* Washington, D.C.: Mathematical Association of America, 1991. "Fifty-five cultural reasons why too few women win at mathematics," successful women mathematicians of various ethnic/racial backgrounds, and much more.

Kogelman, Stanley, and Joseph Warren. *Mind Over Math.* New York: Dial Press, 1978. Exploration of the causes of math anxiety, and suggestions for overcoming the problem.

Langbort, Carol, and Virginia Thompson. *Building Success in Math.* Belmont, Calif.: Wadsworth, 1985. Nontraditional approaches to learning math for people of all ages who feel uncomfortable with the subject.

Pearson, Willie, Jr., and H. Kenneth Bechtel, eds. *Blacks, Science, and American Education.* New Brunswick, N.J.: Rutgers University Press, 1989. Analysis of the past, present, and future of African Americans in mathematics and science, and strategies to insure full participation.

Ruedy, Elisabeth, and Sue Nirenberg. *Where Do I Put the Decimal Point?* New York: Henry Holt, 1990. User-friendly book designed to build the confidence of math fearers through psychological and mathematical methods.

Tobias, Sheila. *Overcoming Math Anxiety.* New York: Norton, 1978. Addressed mainly to middle-class white women, this book has become a classic.

———. *Succeed with Math: Every Student's Guide to Conquering Math Anxiety.* New York: College Board Publication, 1987. Aimed at students entering college; case studies of contemporary applications of mathematics.

RESOURCES

Biography

Bedini, Silvio A. *The Life of Benjamin Banneker.* New York: Scribner, 1972. Life of the eighteenth-century African-American mathematician and astronomer.

Comer, James P. *Maggie's American Dream.* New York: New American Library, 1988. Inspirational story of an uneducated and unlettered African-American woman's fierce determination to educate her four children; they earned thirteen college degrees among them. Told in her own words and in those of her son, a child psychiatrist and associate dean of Yale Medical School.

Grinstein, Louise, and Paul J. Campbell, eds. *Women of Mathematics: A Biographical Sourcebook.* New York: Greenwood Press, 1987. Biographies of forty-three women, from ancient times to the present.

Kenschaft, Patricia C. "Black Women in Mathematics in the United States." *American Mathematical Monthly* 88 (1981): 592–604. Reprinted with additions in *Journal of African Civilization* 4, 1 (1982): 63–83. Brief biographies of women who have earned doctorates in pure mathematics.

———. "Black Men and Women in Mathematics Research." *Journal of Black Studies* 18, 2 (1987): 170–190. Historical and contemporary.

Koblitz, Ann Hibner. *A Convergence of Lives. Sofia Kovalevskaia: Scientist, Writer, Revolutionary.* Boston: Birkhauser, 1983. Exciting biography of a nineteenth-century Russian woman, the first female mathematician in recent times to receive a university appointment; her false marriage, her involvement in revolutionary movements, her close friendships.

Newell, Virginia K., et al. *Black Mathematicians and Their Works.* Ardmore, Pa.: Dorrance, 1980. Brief biographies and bibliographies of the works of sixty-two twentieth-century mathematicians and mathematics educators, as well as eighteenth-century Benjamin Banneker. Many are represented by works of original mathematical research.

Osen, Lynn. *Women in Mathematics.* Cambridge, Mass.: MIT Press, 1974. Biographical sketches of prominent historical figures.

Perl, Teri. *Math Equals.* Menlo Park, Calif.: Addison-Wesley, 1978. Easy-to-read biographies of ten prominent women mathematicians, and activities for the reader to carry out.

———. *Women and Numbers.* San Carlos, Calif.: Wide World Publishing/Tetra, 1993. Biographies of women mathematicians of varied ethnic/racial backgrounds, many of them contemporary; relevant activities; appropriate for young people.

Sands, Aimee. "Never Meant to Survive: A Black Woman's Journey." *Radical Teacher* 30 (1986): 8–15. Computer consultant Evelyn Hammond's encounters with racism and sexism, told in her own words.

Van Sertima, Ivan, ed. *Blacks in Science: Ancient and Modern.* New Brunswick, N.J.: Transaction Books, 1983. Two main themes: scientific and mathematical developments in ancient Africa and contemporary African-American scientists.

Textbooks and Study Materials

Also see titles under "Equity Issues" earlier in this bibliography.

Crandall, JoAnn, et al. *English Language Skills for Basic Algebra.* Englewood Cliffs, N.J.: Prentice Hall, 1987. Appropriate for students whose first language is not English.

Falstein, Lina. *Basic Mathematics: You Can Count on Yourself.* Reading, Mass.: Addison-Wesley, 1986. User-friendly worktext covering arithmetic and elementary algebra, using real-life applications.

Hurwitz, Lucy, and Lou Ferlinger. *Statistics for Social Change.* Boston: South End Press, 1988. Introduction to statistical techniques; their use and misuse in explaining real-life events.

Jacobs, Harold R. *Mathematics: A Human Endeavor. A Book for Those Who Think They Don't Like the Subject.* New York: Freeman, 1982. Appealing introduction to mathematical ways of thinking, algebra, number theory, statistics, and other topics; well illustrated.

Schwartz, Richard H. *Mathematics and Global Survival: Scarcity, Hunger, Population Growth, Pollution, Waste.* Needham Heights, Mass.: Ginn, 1989. Computation, statistics, and graphing applied directly to social issues.

Zaslavsky, Claudia. *Multicultural Mathematics: Interdisciplinary Cooperative Learning Activities.* Portland, Maine: J. Weston Walch, 1993. Historical, cultural, and present-day applications, requiring only a knowledge of arithmetic.

General

Brown, Stephen I., and Marion I. Walter. *The Art of Problem Posing.* Hillside, N.J.: Lawrence Erlbaum, 1983. Raises the question: "What if not?"

Campbell, Douglas M., and John C. Higgins, eds. *Mathematics: People, Problems, Results.* 3 vols. Belmont, Calif.: Wadsworth, 1984. Anthology that gives the nonmathematician "some insight into the nature of mathematics and those who create it."

Davis, Philip J., and Reuben Hersh. *The Mathematical Experience.* Boston: Houghton Mifflin, 1981. Content, history, and philosophy of mathematics, brief biographies, pedagogical issues; well illustrated.

Gardner, Martin. Author of many books of intriguing puzzles and problems.

Kohl, Herbert. *Mathematical Puzzlements: Play and Invention with Mathematics.* New York: Schocken, 1987. Over 250 games, puzzles, paradoxes, and problems, ranging from simple to sophisticated.

Pappas, Theoni. *The Joy of Mathematics.* San Carlos, Calif.: Wide World Publishing, 1989. Math in all spheres of life, delightful reading.

Paulos, John Allen. *Innumeracy: Mathematical Illiteracy and Its Consequences.* New York: Hill and Wang, 1988. Humorous anecdotes point to the social costs of innumeracy on both the personal and national level.

Zaslavsky, Claudia. *Africa Counts: Number and Pattern in African Culture.* New York: Lawrence Hill Books, 1979. Numeration, numbers in daily life, geometry in art and architecture, and mathematical games, illustrated by anecdotes and photographs.

Mainly for Parents

Alper, Lynne, and Meg Homberg. *Parents, Kids, and Computers.* Sybex, 1984 (order from EQUALS; see "Distributors"). The best ways for children to work with computers in school and at home.

AAAS. *Math Power in School, Math Power at Home, Math Power in the Community.* Washington, D.C.: AAAS, 1992 (see "Organizations"). Tools to help all students achieve mathematical power.

Barrata-Lorton, Mary. *Workjobs for Parents.* Menlo Park, Calif.: Addison-Wesley, 1975. Activity-centered learning in the home; lovely photographs.

Burns, Marilyn. *The I Hate Mathematics Book* and several other books for primary and middle-grade students, published by Little, Brown. Lively approach, clever illustrations.

Campbell, Patricia B. *Encouraging Girls in Math and Science.* Newton, Mass.: WEEA Publishing Center, 1992. Four pamphlets, for educators, parents, and community involvement.

Children's Television Workshop, One Lincoln Plaza, New York, NY 10023. Materials for parents in connection with "Square One" and other math and science children's programs.

College Board publishes materials for parents.

EQUALS publishes many activity books addressed to girls, but suitable for all students; career goals are included.

Golick, Margie. *Deal Me In! The Use of Playing Cards in Learning and Thinking.* New York: Norton, 1973. Over a hundred games, diversions, and tricks for children aged four and up, classified by age level.

Grunfeld, Frederick V., ed. *Games of the World.* New York: Ballantine, 1975. Over a hundred games, with cultural background; beautiful illustrations.

Kaye, Peggy. *Games for Math: Playful Ways to Help Your Child Learn Math.* New York: Pantheon, 1987. Resource book for parents of children in grades kindergarten through three.

Papert, Seymour. *Mindstorms: Children, Computers, and Powerful Ideas.* New York: Basic Books, 1980. Background and implications of computers in the lives of children; focus on Papert's computer language LOGO.

Rasmussen, Lori. *Miquon Lab Materials.* Key Curriculum Press, P.O. Box 2304, Berkeley, CA 94702. Use with manipulatives to help children in grades one to three become creative and independent problem solvers. Request catalog.

Sanders, Jo S., and Antonia Stone. *The Neuter Computer: Computers for Girls and Boys.* New York: Neal-Schuman, 1986. Computer activities, equity strategies, and resources to interest girls and boys in computers.

Skolnick, Joan, Carol Langbort, and Lucille Day. *How to Encourage Girls in Math and Science: Strategies for Parents and Educators.* Palo Alto, Calif.: Dale Seymour Publications, 1982. For girls, preschool to grade eight.

Sprung, Barbara, Merle Froschl, and Patricia B. Campbell. *What Will Happen If . . . Young Children and the Scientific Method.* New York: Educational Equity Concepts, 1985. Science activities for elementary school children by educators who value equity.

Stenmark, Jean K., Virginia Thompson, and Ruth Cossey. *Family Math.* Berkeley, Calif.: Lawrence Hall of Science, 1986 (order from EQUALS). Available in Spanish. (See chapter 8 for description.)

Zaslavsky, Claudia. *Preparing Young Children for Math: A Book of Games.* New York: Schocken, 1986. Over a hundred "homemade" activities for children aged two to eight, to encourage parents and teachers to make up their own. Includes games from various cultures.

Notes

Chapter 1. Who's Afraid of Math?

1. Denise Grady, "Can Heart Disease Be Reversed?" *Discover,* March 1987, 55–68.
2. Russell Baker, "Observer: How to Miss Mozart," *New York Times,* 26 October 1991, 19.
3. "Drug May Help the Overanxious on S.A.T.'s," *New York Times,* 22 October 1987, A27.
4. Steve Post, WNYC, New York, 22 October 1987.
5. Marilyn Frankenstein, *Relearning Mathematics* (London: Free Association Books, 1989), 40.
6. Frankenstein, *Relearning,* 11.
7. Judith Weller, personal communication.
8. Deborah H. Hallett, "Remarks for Remediation Panel," *Proceedings of the Fourth International Congress on Mathematical Education* (Boston: Birkhauser, 1983), 656–657.
9. David Henderson, "Mathematics and Liberation," *For the Learning of Mathematics* 1, 3 (March 1981): 12–13.

Chapter 2. Who Needs Math? Everybody!

1. Jeffrey Schmalz, "Cuomo Urges Business Leaders to Aid Minority Youth Training," *New York Times,* 29 April 1987, B1.
2. Patricia C. Kenschaft, "Evelyn Boyd Granville (1924–)," in Louise S. Grinstein and Paul J. Campbell, eds., *Women of Mathematics* (New York: Greenwood Press, 1987), 57–61.
3. Debra K. Rubin, "Fifth Annual Salary Survey," *Working Woman,* January 1984, 59; Robert Pear, "Study Says Affirmative Rule Expands Hiring of Minorities," *New York Times,* 19 June 1983, A16.
4. *New York Times,* 21 September 1990, A1.
5. Notice of Examination, Sanitation Worker Exam. No. 9147, City of New York.
6. "Employment in the Service Industry, Impetus to the 80's Boom, Falters," *New York Times,* 2 January 1992, D4.
7. Doron P. Levin, "Smart Machines, Smart Workers," *New York Times,* 17 October 1988, D1; Michael Schrage, "Innovation: Statistics Skills Would Help U.S. Compete," *Los Angeles Times,* 14 March 1991, D1.

8. "Computers at the School of Art," *Michigan Today,* December 1987, 10.

9. Phyllis Steinmann, "Prescription for Mathematics Anxiety: Some Case Histories," *Focus on Learning Problems in Mathematics* 5, nos. 3 & 4 (Summer/Fall 1983): 21.

10. National Research Council, *Everybody Counts: A Report to the Nation on the Future of Mathematics Education* (Washington, D.C.: National Academy Press, 1989), 53.

11. Steinmann, "Prescription for Mathematics Anxiety," 22.

12. National Research Council, *Everybody Counts,* 32–33.

13. National Science Foundation, *Educating Americans for the 21st Century* (Washington, D.C.: Author, 1983), 13–14.

14. National Science Foundation. *Women and Minorities in Science and Engineering* (Washington, D.C.: Author, 1990), vii–ix.

15. *New York Times,* 4 September 1987, A1.

16. National Science Foundation, *Women and Minorities in Science and Engineering* (Washington, D.C.: Author, 1988), vii–viii.

17. American Association for the Advancement of Science, *Science for All Americans: A Project 2061 Report on Literacy Goals in Science, Mathematics, and Technology* (Washington, D.C.: Author, 1989), 156–157.

Chapter 3. Myths of Innate Inferiority

1. Daniel Goleman, "Girls and Math: Is Biology Really Destiny?" *New York Times,* 2 August 1987, EDUC 42–43.

2. C. P. Benbow and J. C. Stanley, "Sex Differences in Mathematical Ability: Fact or Artifact?" *Science* 210 (12 December 1980): 1264.

3. D. A. Williams and P. King, "Do Males Have a Math Gene?" *Newsweek,* 15 December 1980, 73.

4. "The Gender Factor in Math," *Time,* 15 December 1980, 57.

5. "Are Boys Better at Math?" *New York Times,* 7 December 1980, 102.

6. Alice T. Schafer and Mary W. Gray, "Sex and Mathematics," *Science* 211 (16 January 1981), Editorial; J. E. Jacobs and J. S. Eccles, "Gender Differences in Math Ability: The Impact of Media Reports on Parents," *Educational Researcher* 14, 3 (1985): 20–24.

7. Goleman, "Girls and Math."

8. Lindsay A. Tartre, "Spatial Skills, Gender, and Mathematics," in *Mathematics and Gender,* Elizabeth Fennema and Gilah C. Leder, eds. (New York: Teachers College Press, 1990), 27–59.

9. Roger Sperry, "Some Effects of Disconnecting the Cerebral Hemispheres," *Science* 217 (1982): 1224. See also Joe Alper, "Brain Asymmetries: Sex Roles or Sex Differences?" *Science for the People* 17 (September 1985): 25–29.

10. "Complex Math for a Complex Brain," *Science News* 121 (23 January 1982), 58.

11. Marcia S. Linn and Anne C. Petersen, "Facts and Assumptions about the Nature of Sex Differences," in *Handbook for Achieving Sex Equity through Education,* Susan S. Klein, ed. (Baltimore: Johns Hopkins University Press, 1985), 53–77; Patricia B. Campbell, "Girls and Math: Enough Is Known for Action," *WEEA Digest* (June 1991): 1–3.

12. My discussion of spatial ability relies heavily on Anne C. Petersen and Kathryn E. Hood, "The Role of Experience in Cognitive Performance," in *Women at Work: Socialization toward Inequality,* G. M. Vroman, D. Burnham, and S. G. Gordon, eds. (New York: Gordian Press, 1988), 52–57.

Sandra Blakeslee, in her *New York Times* article, "Female Sex Hormone Is Tied to Ability" (18 November 1988, A1), states that the hormone estrogen may have an effect on the speed with which women can mentally rotate objects drawn on paper. Dr. Doreen Kimura, at the University of Western Ontario, found that when estrogen levels were low, as at the beginning of the menstrual cycle, women could solve more problems within a given time span than when levels were high, in the middle of the cycle. She warned that individuals vary widely, and that one is not likely to encounter such problems in real life, where it's mostly a matter of "finding your car in a parking lot." A note in *ZPG Reporter* (February 1992, page 8) concerns research at the same university indicating that men's spatial ability varies inversely with the level of the sex hormone testosterone, which is generally higher in the fall than in the spring. This could make as much as a 50-point difference in the SAT math score, they say.

13. Joan Ferrini-Mundy, "Spatial Training for Calculus Students: Sex Differences in Achievement and in Visualization Ability," *Journal for Research in Mathematics Education* 18 (1987): 126–140.

14. Janet Elder, "Students' Spatial Skills Decline," *New York Times,* 15 August 1985, C1.

15. Margaret Courtney-Clark, *Ndebele: The Art of an African Tribe* (New York: Rizzoli International Publishers, 1986).

16. David Owen, *None of the Above: Behind the Myth of Scholastic Aptitude* (Boston: Houghton Mifflin, 1985).

17. Owen, *None of the Above;* "SAT Coaching: Reality vs. Cover-up," *FairTest Examiner* 5, 4 (Fall 1991): 1–3.

18. Campbell, "Girls and Math," 2.

19. Sonia Nieto, *Affirming Diversity: The Sociopolitical Context of Multicultural Education* (New York: Longman, 1992), 148–149.

20. Aimee Majoros with Lester Salvador, "Not All Asians Are Math and Science Geniuses," *New Youth Connections,* April/May 1989, 7.

21. "100-Point Lag Found in Blacks' S.A.T. Scores," *New York Times,* 5 October 1982, A21. Between 1976 and 1992, scores for African Americans on the mathematics SAT rose 31 points to 385, still 91 points behind the national norm, according to Karen DeWitt, "Test Scores Are Up Slightly for College-Bound Students," *New York Times,* 27 August

1992, A18. In the past few years blacks have improved their math SAT scores more than any other group.

22. Kenneth B. Clark, "Blacks' S.A.T. Scores," *New York Times,* 21 October 1982, A31.

23. William Glaberson, "U.S. Court Says Awards Based on S.A.T.s Are Unfair to Girls," *New York Times,* 4 February 1989, 1; *The AAUW Report: How Schools Shortchange Girls* (Washington, D.C.: American Association of University Women, 1992), 52–57; Howard Wainer and Linda S. Steinberg, "Sex Differences in Performance on the Mathematics Section of the Scholastic Aptitude Test: A Bidirectional Validity Study," *Harvard Educational Review* 62, 3 (Fall 1992): 323–336.

24. Steven Goldberg, "Numbers Don't Lie," *New York Times,* 5 July 1989, Op-Ed; Arthur M. Jaffe and Arthur S. Wightman, "No Deficit on Campus," *New York Times,* 21 July 1989, Letters.

25. For the history of IQ tests I have relied upon Stephen Jay Gould, *The Mismeasure of Man* (and of woman, I might add) (New York: Norton, 1981).

26. A. R. Jensen, "How Much Can We Boost IQ and Scholastic Achievement?" *Harvard Educational Review* 33 (1969): 1–123.

27. "Born Dumb?" *Newsweek,* March 31, 1969, 84. See, for example, Daniel Goleman, "An Emerging Theory on Blacks' I.Q. Scores," *New York Times,* 10 April 1988, EDUC 22–24.

28. "Irrational Testing + Irrational Licensing = Job Discrimination," *Committee Report* 1, 4 (Fall 1987, Lawyers Committee for Civil Rights Under Law): 3–4; "Testing 'Red Herring' Blocks Civil Rights Bill," *FairTest Examiner* 5, 3 (Summer 1991): 16.

29. See Elaine Mensch and Harry Mensch, *The IQ Mythology: Class, Race, Gender and Inequality* (Carbondale, Ill.: Southern Illinois University Press, 1991).

30. Jonathan Beckwith, "Gender and Math Performance: Does Biology Have Implications for Educational Policy?" *Journal of Education* 105 (1983): 158–174. Also see Katherine Connor and Ellen J. Vargas, *Gender Bias in Standardized Testing,* Proceedings of a Hearing Co-sponsored by the National Commission on Testing and Public Policy and the National Women's Law Center, Washington, D.C., 13 October 1989.

31. Petersen and Hood, "Role of Experience."

32. Eleanor Wilson Orr, *Twice As Less* (New York: Norton, 1987).

33. Reuven Feuerstein, *Instrumental Enrichment* (Baltimore: University Park Press, 1980).

34. Schafer and Gray, "Sex and Mathematics."

35. Beckwith, "Gender and Math Performance."

36. Susan F. Chipman, "Far Too Sexy a Topic," *Educational Researcher* 17, 3 (1988): 49.

37. Arthur B. Powell and Stuart Varden, "Computing Literacy for Working Class Adults," *Interface: The Computer Education Quarterly* 5, 4 (1983–1984): 49.

38. *Newsletter* of Women and Mathematics Education 7 (Fall 1984).

39. As I write this book, the myths of innate inferiority are alive and well. Michael Levin, philosophy professor at the City College of New York, has written articles and is working on a book claiming that blacks on average are less intelligent than whites. Nor are women off the hook, in his estimation. In his antifeminist book, *Feminism and Freedom* (New Brunswick, N.J.: Transaction Publications, 1987), he proclaims women's innate inferiority in mathematics. His wife, a mathematician, is the exception that proves the rule. For Levin's views on women, see Susan Faludi, *Backlash: The Undeclared War Against American Women* (New York: Crown Publishers, 1991), 296–300. For his writings on blacks, see Robert D. McFadden, "Court Finds a Violation of a Professor's Rights," *New York Times,* 9 June 1992, B3.

40. Laurie Hart Reyes and George M. A. Stanic, "Race, Sex, Socioeconomic Status, and Mathematics," *Journal for Research in Mathematics Education* 19 (1988): 27; Campbell, "Girls and Math."

41. Petersen and Hood, "Role of Experience."

42. Johnnetta B. Cole, *Conversations: Straight Talk With America's Sister President* (New York: Doubleday, 1993), 169.

Chapter 4. "A Mind Is A Terrible Thing To Waste": Gender, Race, Ethnicity, and Class

1. "Child Poverty Worsens in 1990," *CDF Reports* 13 (November 1991): 1; "Census Bureau Releases Report on American Indian Tribes," United States Department of Commerce, 7 February 1990. See "Resilience, Schooling, and Development in African-American Youth," the special issue of the journal, *Education and Urban Society* 24 (November 1991).

2. National Science Board, *Educating Americans for the 21st Century* (Washington, D.C.: National Science Foundation, 1983), 13.

3. Doris Entwisle and Karl L. Alexander, "Beginning School Mathematics Competence: Minority and Majority Comparisons," Report No. 33 (March 1989), Center for Research on Elementary and Middle Schools, The Johns Hopkins University. See also Herbert P. Ginsburg and Robert L. Russell, *Social Class and Racial Influence on Early Mathematics Thinking* (Chicago: University of Chicago Press, 1981), 56. Studies in Florida, within Israel, and in the Israeli-occupied West Bank show that socioeconomic status can be a more significant factor than either sex or ethnicity in influencing mathematical performance and attitude toward mathematics, according to Mary W. Gray, "Socioeconomic Factors in Mathematics Achievement," presentation at the Sixth International Congress on Mathematical Education, Budapest (1988). For further confirmation, see *The AAUW Report: How Schools Shortchange Girls* (AAUW Educational Foundation and National Education Association, 1992), 33–34.

4. Isabel Wilkerson, "Class for the Gifted Requires More Than Brains," *New York Times,* 21 February 1990, A1.

5. Don Terry, "Cynicism and Hope: Diverse Voices of Black Youth," *New York Times,* 3 June 1992, A14. See the report, National Research Council, *A Common Destiny: Blacks and American Society* (Washington, D.C.: National Academy Press, 1989), showing that the status of blacks as compared with whites had actually worsened in the previous fifteen years in many areas, including education and earnings.

6. Research conclusions regarding females and mathematics are from Patricia Clark Kenschaft and Sandra Zaroodny Keith, eds., *Winning Women into Mathematics* (Washington, D.C.: Mathematical Association of America, 1991), 11–12, and *The AAUW Report* (1992), 10–13, 28–31, unless other references are given. See also Elizabeth Fennema and Gilah C. Leder, *Mathematics and Gender* (New York: Teachers College Press, 1990), for an excellent analysis of current research.

7. "Math and Career Achievement: A Psychological Model for Decision-Making," *Research News* (University of Michigan) 23 (September–October 1982): 21–22.

8. Phyllis Steinmann, "Prescription for Mathematics Anxiety: Some Case Histories," *Focus on Learning Problems in Mathematics* 5 (Summer/Fall 1983): 21.

9. Susan Gross, *Participation and Performance of Women and Minorities in Mathematics,* Executive Summary (Rockville, Md.: Montgomery County Public Schools, 1988), E7.

10. "Centennial Reflections on Women in American Mathematics," *Newsletter* (Association for Women in Mathematics) 18 (November–December 1988): 8.

11. "Making Everybody Count . . ." *Focus* 12 (April 1992): 10–13.

12. Fred Chichester, personal communication.

13. Frank J. Swetz, "Cross-Cultural Insights into the Question of Male Superiority in Mathematics: Some Malaysian Findings," in *Mathematics, Education, and Society,* Christine Keitel, Peter Damerow, Alan Bishop, and Paulus Gerdes, eds. (Paris: UNESCO, 1989): 139–140.

14. Personal communication.

15. *Women and Minorities in Science and Engineering* (Washington, D.C.: National Science Foundation, 1990), 125.

16. Westina Matthews, "Influences on the Learning and Participation of Minorities in Mathematics," *Journal for Research in Mathematics Education* 15 (1984): 84–95.

17. James P. Comer, "Black Education: A Holistic View," *The Urban Review* 8 (1975): 162–170 (quote p. 167). Also see James P. Comer, "Educating Poor Minority Children," *Scientific American* 259 (November 1988): 42–48; John Ogbu, *Minority Education and Caste* (New York: Academic Press, 1978); Seth Mydans, "Black Identity vs. Success and Seeming 'White,' " *New York Times,* 25 April 1990, B9; Paul Weckstein, "Education and the Economy: Why Better Schools Won't

Necessarily Lead to Better Jobs," *Rethinking Schools* 7, 2 (Winter 1992–1993): 17.

18. Jeff Howard and Ray Hammond, "Rumors of Inferiority: The Hidden Obstacles to Black Success," *New Republic,* 29 September 1985, 17–21; Christina Robb, "Teaching the Basics of Motivation," *Boston Globe,* 12 October 1989, 81.

19. Lori B. Miller, "Black Youths Match Minds in Olympics," *New York Times,* 6 July 1987, 12.

20. Teri Perl, *Women and Numbers* (San Carlos, Calif.: Wide World Publishing/Tetra, 1993): 121–137.

21. William Celis 3d, "Hispanic Dropout Rate Stays High, Since Children Work in Hard Times," *New York Times,* 14 October 1992, B9.

22. Olga Ramirez, "Mathematics Anxiety and Hispanic Undergraduates," *Newsletter* (Women and Mathematics Education) 10 (June 1988): 3, 5.

23. Jaime Escalante and Jack Dirmann, "The Jaime Escalante Math Program," *Journal of Negro Education* 59 (Summer 1990): 407–423; William J. Bennett, *James Madison High School* (Washington, D.C.: United States Department of Education, 1987): 28–29.

24. Douglas Blancero, "Italian-American Dropouts: The Silent Minority," *School Voices* 1 (Summer 1991): 5; Felicia R. Lee, "20% Dropout Rate Found for Italian Americans," *New York Times,* 1 May 1990, B4.

25. "Mastering Mathematics," *Research News* (University of Michigan) 43 (Spring 1992): 2–5.

26. Yuh-Yng Lee, "I'm No Math Whiz," *New Youth Connections* (November 1991): 7.

27. Lucy Sells, "Mathematics—A Critical Filter," *Science Teacher* 75 (1978): 28–29.

28. Personal communication.

29. Quoted with permission, from a letter to Camille Benbow.

30. "Overcoming Disadvantages," *New York Times,* 29 March 1986, 25.

31. Patricia C. Kenschaft, "Black Women in Mathematics in the United States," *American Mathematical Monthly* 88 (1981): 592–604.

32. Marcela Kogan, "Puerto Rican Feminist Speaks," *New Directions for Women* (May/June 1987): 5; Carmen Gautier-Mayoral, "Forests Suffer for the Drug War," *New York Times,* 18 August 1990, 24, Letters.

33. Vera A. Preston and Linda Skinner, "Living within the Circle," *Kui Tatk* 2 (Winter 1986): 6.

34. Sau-Lim Tsang, "The Mathematics Education of Asian Americans," *Journal for Research in Mathematics Education* 15 (1984): 114–122.

35. *Take It from Us . . . You Can Be an Engineer* (Fairfield, Conn.: General Electric Educational Communications Programs, 1987), unpaged.

36. Horace B. Deets, "Older Americans Go 'High Tech' Despite Ads," *AARP Bulletin* 32 (May 1991): 3.

37. Teri H. Perl, *Math Equals* (Menlo Park, Calif.: Addison-Wesley, 1978), 101–111; Lord Vivian Bowden, "The Language of Computers," in

Mathematics: People, Problems, Results, Douglas M. Campbell and John C. Higgins, eds. (Belmont, Calif.: Wadsworth International, 1984), vol. 3, 2–14.

38. Claudia Zaslavsky, "Who Invented COBOL?" *Newsletter* of the Association for Women in Mathematics 19 (January 1989): 3–5; Jean E. Sammet, "Farewell to Grace Hopper—End of an Era!" *Communications of the ACM* 35 (April 1992): 128–131.

Chapter 5. Our Schools Are Found Wanting

1. National Commission on Secondary Schooling for Hispanics, *Make Something Happen—Hispanics and Urban High School Reform* (Washington, D.C.: Hispanic Policy Development Project, 1984). The Hispanic dropout rate has remained fairly stable since the mid-1970s, according to *The Condition of Education 1992* (Washington, D.C.: National Center for Educational Statistics, 1992), 7, a statistic that should cause alarm.

2. U.S. Department of Education, *What Works. Schools That Work: Educating Disadvantaged Children* (Pueblo, Colo.: Schools That Work, 1987), 3.

3. National Science Board Commission on Precollege Education in Mathematics, Science and Technology, *Educating Americans for the 21st Century* (Washington, D.C.: National Science Foundation, 1983), vii.

4. Speech quoted in *New York Times,* 30 October 1984, B4.

5. Carnegie Foundation for the Advancement of Teaching, *An Imperiled Generation: Saving Urban Schools* (Princeton, N.J.: Princeton University Press, 1988).

6. Jane Perlez, "School Basks in Spotlight of Contest," *New York Times,* 14 January 1988, B1. In District 4, in East Harlem, students may choose to attend any junior high school; each school features a different specialization. The district also supports the Central Park East alternative elementary and secondary schools, affiliated with Theodore Sizer's Coalition of Essential Schools; see Deborah Meier, "Success in East Harlem," *American Educator* 11 (Fall 1987): 34–39.

7. Doxey Wilkerson, *Special Problems of Negro Education* (Washington, D.C.: U.S. Government Printing Office, 1939), discussed in Maceo Crenshaw Dailey Jr. and Ernest D. Washington, "The Evolution of Doxey A. Wilkerson, 1935–1945," *Freedomways* 25 (Summer 1985): 101–115. See H. Kenneth Bechtel's Introduction, in Willie Pearson, Jr., and H. Kenneth Bechtel, eds., *Blacks, Science, and American Education* (New Brunswick, N.J.: Rutgers University Press, 1989), 1–10, for a history of African-American education.

8. Ann Bastian et al., *Choosing Equality: The Case for Democratic Schooling* (Philadelphia: Temple University Press, 1986), 47; Karen De Witt, "Rising Segregation Is Found for Hispanic Students," *New York Times,* 9 January 1992, A15.

9. See Conrad K. Harper and Stuart Land, "The Injustice Department," *New York Times,* 14 April 1988, Op-Ed. The authors objected to the dismissal by the U.S. Department of Justice of over two hundred school desegregation cases in Georgia, Alabama, and Mississippi.

10. Fred Hechinger, "About Education: Debate on the Role of Elite Schools," *New York Times,* 5 February 1982, C8; Patricia C. Kenschaft, "Black Women in Mathematics in the United States," *American Mathematical Monthly* 88 (1981): 592–604.

11. William J. Bennett, *James Madison High School: A Curriculum for American Students* (Washington, D.C.: U.S. Department of Education, 1987), 16–17; television station WWOR documentary, broadcast 27 November 1988.

12. Lorraine Hansberry, "The Scars of the Ghetto," *Monthly Review* 41 (July/August 1989): 52–55.

13. Amy Stuart Wells, "Asking What Schools Have Done, Or Can Do, to Help Desegregation," *New York Times,* 16 January 1991, B6.

14. Jane Perlez, "New York School Buildings Scarred by Years of Neglect," *New York Times,* 13 May 1987, A1. Also see Susan Breslin, *Promoting Poverty in New York City Schools* (New York: Community Service Society, 1987), for inequities among local schools.

15. Jonathan Kozol, *Savage Inequalities: Children in America's Schools* (New York: Crown Publishers, 1991). See p. 16 for the story of chemical companies and East St. Louis. On pp. 223–229 Kozol discusses the inequity in Texas school funding in 1991, ranging from $2,000 to $19,000 per student. A later report in the *New York Times* (4 December 1992, A26) describes per pupil spending in Texas as ranging from $2,337 to $56,791! California equalizes per pupil spending in all its districts, but at a rate that is among the lowest in the nation. In 1978 California voters carried out a tax revolt with a vengeance when they voted for Proposition 13, a measure to limit property taxes. Other states followed suit soon afterwards. In wealthy areas parents raise money for school necessities, but poor schools must make do with what they have.

16. Deborah A. Verstegen and David L. Clark, "The Diminution in Federal Expenditures for Education during the Reagan Administration," *Phi Delta Kappan* 70 (1988): 134–138; Margaret Spillane and Bruce Shapiro, "A Small Circle of Friends," *The Nation,* 21 September 1992, 281. William Celis 3rd, in his article, "A Texas-Size Battle to Teach Rich and Poor Alike," *New York Times,* 12 February 1992, A23, notes that Texas and twenty-two other states are in court over the way they finance public schools. On the same page are listed the per pupil expenditures in several New York State districts in the 1990–1991 school year: $6,644 in New York City and about $30,000 in some suburban Long Island towns. In 1981 a lawsuit was brought in the New Jersey courts challenging inequitable school funding. The New Jersey Supreme Court ruled in 1990 that the state had the obligation to equalize funding among the poorest cities and the wealthiest suburbs,

noting that Princeton schools had one computer for every eight students, while the ratio in Camden was one computer for fifty-eight students, and Paterson schools limited computer use to the basic skills remedial program mandated by the federal government. The court concluded: "We find that under the present system the evidence compels but one conclusion: the poorer the district and the greater its need, the less the money available, and the worse the education. . . . Such funding must be guaranteed and mandated by the State." The attempt by Democratic Governor James Florio to implement the ruling by imposing additional taxes resulted in an overwhelming victory for the Republicans in the following state election and drastic cuts in the new tax law. See "Excerpts from New Jersey Court Ruling on School Financing," *New York Times,* 6 June 1990, B4.

17. Gene Maeroff, *Don't Blame the Kids* (New York: McGraw-Hill, 1982), 192–197.

18. Lisa Syron, *Discarded Minds: How Gender, Race and Class Biases Prevent Young Women from Obtaining an Adequate Math and Science Education in New York City Public Schools* (New York: Center for Public Advocacy Research, 1987), 35, 86.

19. Susan Gross, *Participation and Performance of Women and Minorities in Mathematics,* Executive Summary (Rockville, Md.: Montgomery County Public Schools, 1988), E4–E12. See Josephine D. Davis, "The Mathematics Education of Black High School Students," in Pearson and Bechtel, eds., *Blacks, Science,* 23–42.

20. Cora Bagley Marrett and Harold Gates, "Male-Female Enrollment across Mathematics Tracks in Predominantly Black High Schools," *Journal for Research in Mathematics Education* 14 (1983): 113–118. In Selma, Alabama, parents and students of the African-American community organized to overturn the school administration's practice of placing half of all African-American students in the lowest track, based on such flimsy criteria as "teacher recommendation" and "student maturity," thus shutting them out from taking algebra and other academic courses. See Stan Karp, "Selma Students Tied to the Track," *Rethinking Schools* 5 (October/November 1990): 1, 10.

21. Leonard A. Valverde, "Underachievement and Underrepresentation of Hispanics in Mathematics and Mathematics-Related Careers," *Journal for Research in Mathematics Education* 15 (1984): 123–133.

22. Elvira Valenzuela Crocker, "The Report Card on Educating Hispanic Women" (Washington, D.C.: Project on Equal Education Rights [PEER], NOW Legal Defense and Education Fund, 1983).

23. Cicely A. Rodway, "I Was Lucky If They Knew My Name: First-Generation Immigrant Students Speak Out," *School Voices* 2 (March 1992): 16.

24. Michael Cole and Peg Griffin, eds., *Contextual Factors in Education* (Madison, Wisc.: Wisconsin Center for Education Research, 1987), 27–28. For a full discussion of tracking, see Jeannie Oakes, *Keeping*

Track: How Schools Structure Inequality (New Haven, Conn.: Yale University Press, 1984), and Jeannie Oakes, *Multiplying Inequalities: The Effects of Race, Social Class, and Tracking on Opportunities to Learn Mathematics and Science* (Santa Monica, Calif.: Rand Corporation, 1990).

25. Edward DeAvila, "Bilingualism, Cognitive Functioning, and Language Minority Group Membership," in *Linguistic and Cultural Influences on Learning Mathematics,* Rodney R. Cocking and Jose P. Mestre, eds. (Hillsdale, N.J.: Lawrence Erlbaum Associates, 1988), 101–121.

26. Edward B. Fiske, "America's Test Mania," *New York Times* (10 April 1988), EDUC 16–20. Also see National Council of Teachers of Mathematics, *Curriculum and Evaluation Standards for School Mathematics* (Reston, Va.: National Council of Teachers of Mathematics, 1989); Susan Chira, "Study Finds Standardized Tests May Hurt Education Efforts," *New York Times,* 16 October 1992, A19; and publications of the National Center for Fair and Open Testing (FairTest).

Other countries are not so caught up in the testing mania. Europeans expect students to solve problems rather than to pick one right answer out of four or five, and teachers are trusted to use their own methods of assessment. In a joint statement, published in *Mathematics Teaching* 122 (March 1988): 50–51, as well as in other journals, several associations of British educators protested a proposed national testing program, citing some of the objections discussed here.

27. Personal communication.

28. Myra and David Sadker, "Sexism in the Classroom: From Grade School to Graduate School," *Phi Delta Kappan* 67 (1986): 512–515; Katha Pollitt, "Coeducation and Adolescent Girls," *New York Times,* 19 December 1985, C2. Cornelius Riordan, in his book, *Girls and Boys in School* (New York: Teachers College Press, 1990), argues that adolescent girls might benefit from single-sex schooling.

29. Patricia Kenschaft, personal communication. Upon hearing about the incident from her daughter, Dr. Kenschaft anonymously sent this teacher a subscription to the newsletter of the Association of Women in Mathematics. Not only did his own classroom behavior change dramatically, but he began to evangelize others.

30. Patricia Lund Casserly, *Helping Able Young Women Take Math and Science Seriously in School* (New York: The College Board, 1979).

31. Theodore Sizer, *Horace's Compromise: The Dilemma of the American High School* (Boston: Houghton Mifflin, 1984), 37.

32. Mina Choi was a finalist in the 1988 Westinghouse Talent Search with her project, "Which of These Students Will Be Most Successful?"

33. "Fewer Degrees for Black Men in Maryland," *New York Times,* 16 November 1988, B8. Also see Diane Scott-Jones and Maxine L. Clark, "The School Experiences of Black Girls: The Interaction of Gender, Race, and Socioeconomic Status," *Phi Delta Kappan* 67 (March 1986): 520–526.

34. Susan Gross, *Participation and Performance.*

35. See, for example, Henry Giroux, *Ideology, Culture, and the Process of Schooling* (Philadelphia: Temple University Press, 1981); Michael Apple and Lois Weis, *Ideology and Practice in Schooling* (Philadelphia: Temple University Press, 1977), as well as their later writings. As applied to mathematics education, see Michael W. Apple, "Do the Standards Go Far Enough? Power, Policy, and Practice in Mathematics Education," *Journal for Research in Mathematics Education* 23, 5 (1992): 412–431; Patricia B. Campbell, "So What Do We Do with the Poor, Non-White Female? Issues of Gender, Race, and Social Class in Mathematics and Equity," *Peabody Journal of Education* 66, 2 (1989): 95–112; Walter G. Secada, "Agenda Setting, Enlightened Self-Interest, and Equity in Mathematics Education," *Peabody Journal of Education* 66, 2 (1989): 22–56.

36. Jean Anyon, "Social Class and the Hidden Curriculum of Work," *Journal of Education* 162 (Winter 1980): 67–92.

37. Herbert Kohl writes: "To agree to learn from a stranger who does not respect your integrity causes a major loss of self. The only alternative is to not-learn and reject the stranger's world." Quoted from *I Won't Learn from You: The Role of Assent in Learning* (Minneapolis: Milkweed Editions, 1991), 16. The students may have felt that the material was not worth learning because it would not lead to anything worthwhile in their lives.

38. "The children of the poor, the underclass, the disadvantaged, the forgotten are educated to become adults who are also members of the underclass—the disadvantaged and the forgotten." Quoted from Ralph Parish et al., "Knock at Any School," *Phi Delta Kappan* 70 (1989): 387.

39. Curtis C. McKnight et al., *The Underachieving Curriculum: Assessing U.S. School Mathematics from an International Perspective* (Champaign, Ill.: Stipes Publishing Company, 1987). See also Jeannie Oakes, *Multiplying Inequalities.* In 1990 The College Board initiated a project, Equity 2000, with the goal: "That by the end of the twentieth century, students from minority and disadvantaged groups will attend and complete college at the same rate as students from non-minority, advantaged groups." To achieve that goal, all students will have studied algebra and geometry by the end of the tenth grade of high school. In a letter to the *New York Times* (26 November 1992), the president of The College Board, Donald M. Stewart, wrote: "Given proper preparation, appropriate support and adequate motivation, all students can learn and succeed in demanding academic courses of the kind needed for college prep or today's changing and challenging world of work."

40. Gerald Coles, *The Learning Mystique: A Critical Look at Learning Disabilities* (New York: Pantheon, 1987), 205.

41. Barbara S. Allardice and Herbert P. Ginsburg, "Children's Psychological Difficulties in Mathematics," in *The Development of Mathematical Thinking* (New York: Academic Press, 1983), 330.

42. Allardice and Ginsburg, "Children's Psychological Difficulties," 348–349.

43. Sheryl Hovey, "College Degree Programs for Working Adults," *On Campus* 7 (May 1988): 8–9.

44. "Democracy's Colleges," *ETS Policy Notes* 3 (Winter 1990): 1, 6–7.

45. Roberta M. Hall and Bernice R. Sandler, *The Classroom Climate: A Chilly One for Women?* (Washington, D.C.: Project on the Status and Education of Women, Association of American Colleges, 1982).

46. Edward B. Fiske, "Even at a Former Women's College, Male Students Are Taken More Seriously, a Researcher Finds," *New York Times,* 11 November 1990, B8.

47. Shiela M. Strauss and Rena F. Subotnik, "Gender Differences in Behavior and Achievement: A True Experiment Involving Random Assignment to Single Sex and Coeducational Advanced Placement (BC) Calculus Classes," Final Report, 1991 (unpublished).

48. Ruth A. Schmidt, "American Women's Colleges Need Not Go Coed or Go Under," *New York Times,* 4 June 1987, A26 Letters.

49. Sylvia T. Bozeman, "Black Women Mathematicians: In Short Supply," *Sage* 6 (Fall 1989): 18–23.

50. Kathleen Teltsch, "U.S. Indian Colleges Get First Big Corporate Gift," *New York Times,* 17 July 1991, A19; William Celis 3rd, "Texas College System Awaits Ruling in Hispanic Bias Case," *New York Times,* 4 December 1991, B16.

51. The College Board, *Equality and Excellence: The Educational Status of Black Americans* (New York: College Entrance Examination Board, 1985), 17–18. See also Laura I. Rendon and Estrella M. Triana, *Making Mathematics and Science Work for Hispanics* (Washington, D.C.: American Association for the Advancement of Science, 1989), and the Quality Education for Minorities Project, *Education That Works: An Action Plan for the Education of Minorities* (Cambridge, Mass.: Massachusetts Institute of Technology, 1990).

52. J. W. Carmichael, Jr., and John P. Sevenair, "Preparing Minorities for Science Careers," *Issues in Science and Technology* 7 (Spring 1991): 55–60.

53. Peter Applebome, "Epilogue to Integration Fight: Blacks Favor Own Colleges," *New York Times,* 29 May 1991, A1, A21; Susan Chira, "[Supreme Court] Ruling May Force Changes at Southern Colleges," *New York Times,* 27 June 1992, 10.

54. Isabel Wilkerson, "Racial Harassment Altering Blacks' Choices on Colleges," *New York Times,* 9 May 1990, A1, B10.

55. Gary Orfield, "Money, Equity, and College Access," *Harvard Educational Review* 62, 3 (1992): 337–372. See Gail E. Thomas, "Black Science Majors in Colleges and Universities," in Pearson and Bechtel, eds., *Blacks, Science,* 59–78.

56. Marsha Lakes Matyas and Shirley M. Malcom, eds., *Investing in Human Potential: Science and Engineering at the Crossroads,* Executive

Summary (Washington, D.C.: American Association for the Advancement of Science, 1991), 9–10. See relevant chapters in Pearson and Bechtel, eds., *Blacks, Science.*

Chapter 6. School Math is Not Necessarily Real Math

1. Hassler Whitney, "Coming Alive in School Mathematics and Beyond," *Educational Studies in Mathematics* 18 (1987): 229–242; Fred M. Hechinger, "Learning Math by Thinking," *New York Times,* 10 June 1986, C1, C7.

2. Judith Richards teaches at the Saundra Graham and Rosa Parks Alternative School in Cambridge, Massachusetts. See Claudia Zaslavsky, "People Who Live in Round Houses," *Arithmetic Teacher* 37 (September 1989): 18–21.

3. Erna Yackel, Paul Cobb, Terry Wood, Grayson Wheatley, and Graceann Merkel, "The Importance of Social Interaction in Children's Construction of Mathematical Knowledge," *Teaching and Learning Mathematics in the 1990s, 1990 Yearbook,* Thomas J. Cooney, ed. (Reston, Va.: National Council of Teachers of Mathematics, 1990), 12–21.

4. Penelope L. Peterson, Elizabeth Fennema, and Thomas Carpenter, "Using Knowledge of How Students Think about Mathematics," *Educational Leadership* 46 (December 1988/January 1989): 42–46.

5. Phyllis Steinmann, "Prescription for Mathematics Anxiety: Some Case Histories," *Focus on Learning Problems in Mathematics* 5 (Summer/Fall 1983): 15–24 (quotes are on 9–10).

6. Alan Schoenfeld, "When Good Teaching Leads to Bad Results," *Educational Psychologist* 23 (1988): 143–166.

7. Cynthia Sylva and Robert P. Moses, "The Algebra Project: Making Middle School Mathematics Count," *Journal of Negro Education* 59 (1990): 375–391.

8. John D. Volmink, "Acquisition of Concepts and Construction of Meaning in Geometry," unpublished manuscript (1987).

9. John Poland, "A Modern Fairy Tale," *American Mathematical Monthly* 94 (1987): 291–295; Gloria F. Gilmer and Scott W. Williams, "An Interview with Clarence Stephens," *UME Trends* 2 (March 1990): 1, 4, 7. The coauthors were students of Stephens at Morgan State University. New York University professor Peter D. Lax begins his article "Calculus Reform: A Modest Proposal," *UME Trends* 2 (May 1990): 1, 4, with the statement: "Among our many educational crimes, the one that bears the greatest responsibility for the shortage of young Americans choosing mathematics, physics, and engineering as careers may very well be the wretched way we teach calculus."

10. AAUW Report, *How Schools Shortchange Girls* (American Association of University Women Educational Foundation, 1992), 29; Alice Miller, "Eureka! Summer Program in Math and Sports for Teenage Girls," *Newsletter* (Association for Women in Mathematics) 18 (July–

August 1988): 22–23. Another program that has been successful in engaging low-income and minority girls in science and math in many parts of the United States is Girls Incorporated's Operation SMART. Plans are underway to combine the Eureka and SMART programs. See the "Resources" section for more information.

11. *Curriculum and Evaluation Standards for School Mathematics* (Reston, Va.: National Council of Teachers of Mathematics, 1989). Relevant reports from the Mathematical Sciences Education Board of the National Research Council are *Everybody Counts* (1989), *Reshaping School Mathematics* (1990), and *Moving beyond Myths: Revitalizing Undergraduate Mathematics* (1991).

12. "Chicago Provides Free Calculators to Students," *New York Times,* 5 January 1988, C10. Calculators have become quite sophisticated; they now do much more than add, subtract, multiply, and divide.

13. I taught in Greenburgh Central School District Seven, just north of New York City. For a discussion of the mathematics curriculum we developed, see Claudia Zaslavsky, "Multicultural Mathematics: One Road to the Goal of Mathematics for All," in *Reaching All Students with Mathematics,* G. Cuevas and M. Driscoll, eds. (Reston, Va.: National Council of Teachers of Mathematics, 1993).

14. See James A. Banks, "The Transformative Approach to Curriculum Reform: Issues and Examples," in *Empowerment through Multicultural Education,* Christine E. Sleeter, ed. (Albany: State University of New York Press, 1991), 125–141. Two texts for adults that incorporate these perspectives are Marilyn Frankenstein, *Relearning Mathematics* (London: Free Association Books, 1989), and Richard H. Schwartz, *Mathematics and Global Survival* (Needham Heights, Mass.: Ginn, 1989).

15. Personal communication. D'Ambrosio was the founder in 1985 of the International Study Group on Ethnomathematics; Dr. Gloria Gilmer is president of the organization.

16. Paulus Gerdes, "How to Recognize Hidden Geometrical Thinking: A Contribution to the Development of Anthropological Mathematics," *For the Learning of Mathematics* 6 (June 1986): 10–12.

17. Books on this subject include Marcia Ascher, *Ethnomathematics: A Multicultural View of Mathematical Ideas* (Pacific Grove, Calif.: Brooks/Cole, 1991); Michael P. Closs, ed., *Native American Mathematics* (Austin: University of Texas Press, 1986); George G. Joseph, *Crest of the Peacock: The Non-European Roots of Mathematics* (New York: St. Martin's Press, 1991); Ivan Van Sertima, *Blacks in Science* (New Brunswick, N.J.: Transaction Books, 1983); Claudia Zaslavsky, *Africa Counts: Number and Pattern in African Culture* (New York: Lawrence Hill Books, 1979).

18. For example, articles by Claudia Zaslavsky: "Bringing the World into the Math Class," *Curriculum Review* 24 (January/February 1985): 62–65, and "Multicultural Mathematics Education for the Middle Grades," *Arithmetic Teacher* 38 (February 1991): 8–13; also Gloria Gilmer, Mary M. Soniat-Thompson, and Claudia Zaslavsky, *Building Bridges*

to *Mathematics: Cultural Connections* (Menlo Park, Calif.: Addison-Wesley, 1992), activity cards for elementary and middle grades.

19. Quoted in Claudia Zaslavsky, "What Is Math For?" *Urban Review* 8 (1975): 232–240.

20. Allyn Jackson, "Multiculturalism in Mathematics: Historical Truth or Political Correctness?" in *Heeding the Call for Change: Suggestions for Curriculum Action,* Lynn Arthur Steen, ed. (Washington, D.C.: Mathematical Association of America, 1992): 121–134 (citation on 130).

21. Mary Harris, "An Example of Traditional Women's Work As a Mathematics Resource," *For the Learning of Mathematics* 7 (November 1987): 26–28.

22. See, for example, Claudia Zaslavsky, "Symmetry in American Folk Art," *Arithmetic Teacher* 38 (September 1990): 6–12, an illustrated article on mathematical symmetry in Dineh (Navajo) rugs and American quilts, and Dorothy K. Washburn and Donald W. Crowe, *Symmetries of Culture* (Seattle: University of Washington Press, 1988). The reader might be surprised to find two articles on symmetry in Hungarian folk needlework in the *Journal of Chemical Education.*

23. Allyn Jackson, "Multiculturalism," 128–129. See also Gloria F. Gilmer, "An Interview with Abdulalim Abdullah Shabazz," *UME Trends* 3 (January 1992): 8.

24. Dr. Carl Downing, director, Increasing the Participation of Native American Students in Higher Mathematics, Central State University, Edmond, Oklahoma. See also Lyn Taylor, Ellen Stevens, John J. Peregoy, and Barbara Bath, "American Indians, Mathematical Attitudes, and the *Standards,*" *Arithmetic Teacher* 38 (February 1991): 14–21.

25. Allyn Jackson, "Multiculturalism," 127–128.

26. L. P. Benezet, "The Teaching of Arithmetic: The Story of an Experiment," *Journal of the National Education Association* 24, no. 8 (1935): 241–244, no. 9 (1935): 301–303, and 25, no. 1 (1936): 2–3; reprinted in the *Humanistic Mathematics Network Newsletter* 6 (May 1991): 2–14.

27. Whitney, "Coming Alive."

Chapter 7. Everybody Can Do Math: Solving the Problem

1. Munir Fasheh, "Mathematics in a Social Context," in *Mathematics, Education, and Society,* Christine Keitel, Peter Damerow, Alan Bishop, and Paulus Gerdes, eds. (Paris: UNESCO, 1989), 84–85. British educator Mary Harris has written from the feminist point of view about the mathematics inherent in various forms of work; see Mary Harris, *Schools, Mathematics and Work* (London: Falmer Press, 1991).

2. Jean Lave, *Cognition in Practice: Mind, Mathematics, and Culture in Everyday Life* (New York: Cambridge University Press, 1988): 47–59.

3. Eric Berger, *Making It in Engineering* (New York: Engineers Council for Professional Development, 1975), 2–3.

4. Patricia Keegan, "Playing Favorites," *New York Times,* 6 August 1989, EDUC 26–27.

5. For suggested readings, see chapter 6, note 17.

6. Sonia Nieto, *Affirming Diversity: The Sociopolitical Context of Multicultural Education* (New York: Longman, 1992): 119; see chapter 5, "Cultural Issues and Their Impact on Learning," 109–152. A useful reference for applications to the learning of mathematics is *Linguistic and Cultural Influences on Learning Mathematics,* Rodney R. Cocking and Jose P. Mestre, eds. (Hillsdale, N.J.: Lawrence Erlbaum Associates, 1988).

7. Floy C. Pepper, "Is There an Indian Learning Style?" *Kui Tatk* 2 (Winter 1986), Native American Science Education Association.

8. Elizabeth Fennema and Gilah C. Leder, eds., *Mathematics and Gender* (New York: Teachers College Press, 1990), 91–92.

9. See, for example, Mary F. Belenky, Blythe M. Clinchy, Nancy Goldberger, and Jill M. Tarule, *Women's Ways of Knowing: The Development of Self, Body, and Mind* (New York: Basic Books, 1986); Carol Gilligan, *In a Different Voice: Psychological Theory and Women's Development* (Cambridge, Mass.: Harvard University Press, 1982); Sue V. Rosser, *Female Friendly Science: Applying Women's Studies Methods and Theories to Attract Students* (New York: Pergamon Press, 1990).

10. Deborah Tannen, "Language, Gender, and Teaching," *Rethinking Schools* 6, 4 (May/June 1992): 7, condensed from her article in the *Chronicle of Higher Education,* 19 June 1991.

11. Gloria Steinem, *Revolution from Within: A Book of Self-Esteem* (Boston: Little, Brown and Company, 1992), 189.

12. Rosser, *Female Friendly Science,* 29–30.

13. Dorothy Buerk, "The Voices of Women Making Meaning in Mathematics," *Journal of Education* 167, 3 (1985): 59–70.

14. Rose Asera, personal communication; Allyn Jackson, "Minorities in Mathematics: A Focus on Excellence, Not Remediation," *American Educator* 13, 1 (Spring 1989): 22–27; "Novel Math Workshops Boost Minority Students," *New York Times,* 5 August 1992, B9.

Similar programs initiated by Dr. Clarence Stephens at Potsdam College, State University of New York, and described in this book in chapter 6, have attracted less publicity. For further discussion of Stephens's work, in addition to the references in chapter 6, note 9, see Pat Rogers, "Student-Sensitive Teaching at the Tertiary Level: A Case Study," *Proceedings of the Twelfth Annual Conference,* 20–25 July 1988, International Group for the Psychology of Mathematics Education, vol. 2, 536–543.

15. Among the books and articles I have used are the following: Dorothy Buerk, "The Voices of Women"; Dorothy Buerk, "Writing in Mathematics: A Vehicle for Development and Empowerment," in *Using Writing to Teach Mathematics,* Andrew Sterrett, ed., MAA Notes #16, 78–84 (Washington, D.C.: The Mathematical Association of America, 1990);

Bonnie Donady and Susan Auslander, "The Math Anxiety Workshop," *Focus on Learning Problems in Mathematics* 1, 4 (October 1979): 57–66; Marilyn Frankenstein, *Relearning Mathematics* (London: Free Association Books, 1989); Allyn Jackson, "Minorities in Mathematics"; Stanley Kogelman, Susan Forman, and Jan Asch, "Math Anxiety: Help for Minority Students," *American Educator* 5 (Fall 1981): 30–32; Stanley Kogelman and Joseph Warren, *Mind Over Math* (New York: McGraw-Hill, 1978); C. Ann Oxrieder and Janet P. Ray, *Your Number's Up* (Reading, Mass.: Addison-Wesley, 1982); Arthur B. Powell and Jose Lopez, "Writing As a Vehicle to Learn Mathematics: A Case Study," in *The Role of Writing in Learning Mathematics and Science,* P. Connolly and T. Vilardi, eds. (New York: Teachers College Press, 1989), 269–303; Elisabeth Ruedy and Sue Nirenberg, *Where Do I Put the Decimal Point?* (New York: Henry Holt and Company, 1990); Phyllis Steinmann, "Prescription for Mathematics Anxiety: Some Case Studies," *Focus on Learning Problems in Mathematics* 5 (Summer/Fall 1983): 15–24; Sheila Tobias, *Overcoming Math Anxiety* (New York: Norton, 1978).

16. Ruedy and Nirenberg, *Where Do I Put . . .* , chapter 6, "Confidence Building Techniques."

17. National Research Council, *Everybody Counts: A Report on the Future of Mathematics Education* (Washington, D.C.: National Academy Press, 1989), 10.

18. Steinmann's Algebra Blox are distributed by the EXA Company, 6928 East Sunnyvale Road, Paradise Valley, AZ 85253. See Leah P. McCoy, "Correlates of Mathematics Anxiety," *Focus on Learning Problems in Mathematics* 14, 2 (1992): 51–57. McCoy writes that the "use of manipulative materials for mathematics instruction has also been found to reduce mathematics anxiety" (p. 51), in particular for the remediation of adults who prefer a tactile-kinesthetic mode of learning.

19. Ruedy and Nirenberg, *Where Do I Put . . .* , 2.

20. Stephen I. Brown and Marion I. Walter, *The Art of Problem Posing,* 2d ed. (Hillsdale, N.J.: Lawrence Erlbaum Associates, 1990).

21. Steinmann, "Prescription," 12–13.

22. Buerk, "Voices," 66–67.

23. Buerk, "Writing," 78.

24. Ruedy and Nirenberg, *Where Do I Put . . .* , 217.

Chapter 8. Families, the First Teachers

1. Thomas H. Maugh II, "Infants Have Math Ability at 5 Months, Study Shows," *Los Angeles Times,* 27 August 1992, A3.

2. Patricia Kenschaft, personal communication.

3. Susan Gross, *Participation and Performance of Women and Minorities in Mathematics,* Executive Summary (Rockville, Md.: Montgomery County Public Schools, 1988), E17.

4. Greenberg-Lake, Inc., *Shortchanging Girls, Shortchanging Amer-*

ica (Washington, D.C.: American Association of University Women, 1991). The results of "a nationwide poll to assess self esteem, educational experience, interest in math and science, and career aspirations of girls and boys ages 9–15."

5. Cited in Sonia Nieto, *Affirming Diversity: The Sociopolitical Context of Multicultural Education* (New York: Longman, 1992), 201.

6. Research Report: "Herbert Ginsburg Looks at 'Personality and Cognition,' " *Teachers College Today* 15, 2 (Fall/Winter 1987).

7. Harold W. Stevenson, "The Asian Advantage: The Case of Mathematics," *American Educator* 11, 2 (1987): 26–31, 47; Harold W. Stevenson and Shin-ying Lee, *Contexts of Achievement: A Study of American, Chinese, and Japanese Children* (Chicago: University of Chicago Press, 1990).

8. Herbert Ginsburg, *Children's Arithmetic: The Learning Process* (New York: Van Nostrand, 1977), 19. See, for example, Claudia Zaslavsky, *Preparing Young Children for Math: A Book of Games* (New York: Schocken Books, 1986), activities for children aged two to eight.

9. See, for example, Constance Kamii, *Young Children Reinvent Arithmetic* (New York: Teachers College Press, 1985). In her introduction, Kamii writes: "Traditional education unwittingly aims at blind obedience rather than critical, independent thinking" (pp. ix–x).

10. Fred M. Hechinger, "About Education: Repeating Kindergarten: Does It Hurt More Than It Helps?" *New York Times,* 14 September 1988, B9. Hechinger is commenting on the report by Lorrie A. Shepard and Mary Lee Smith, "Flunking Kindergarten: Escalating Curriculum Leaves Many Behind," *American Educator* 12, 2 (1988): 34–38. The authors tell of "the principal who visits each May, test scores in hand, seeking an explanation as to why several of the children are not above the national norms." See also David Elkind, *Miseducation: Preschoolers at Risk* (New York: Alfred A. Knopf, 1987).

11. Phyllis Steinmann, "Prescription for Mathematics Anxiety: Some Case Histories," *Focus on Learning Problems in Mathematics* 5, nos. 3 & 4 (1983): 3–4.

12. Virginia Thompson, "FAMILY MATH: Linking Home and Mathematics," in *Mathematics, Education, and Society,* Christine Keitel, Peter Damerow, Alan Bishop, and Paulus Gerdes, eds. (Paris: UNESCO, 1989), 62–65. See also Virginia Thompson, "Family Math," *Mathematics Teacher* 80, 1 (1987): 76. To order the *Family Math* curriculum guide and filmstrip, contact EQUALS, Lawrence Hall of Science, University of California, Berkeley, CA 94720.

13. Cynthia M. Silva and Robert P. Moses, "The Algebra Project: Making Middle School Mathematics Count," *Journal of Negro Education* 59, 3 (1990): 390.

14. J. W. Carmichael, Jr., and John P. Sevenair, "Preparing Minorities for Science Careers," *Issues in Science and Technology* 7, 3 (1991): 55–60; Beatriz C. Clewell, "Intervention Programs: Three Case Studies," in

Willie Pearson, Jr., and H. Kenneth Bechtel, eds., *Blacks, Science, and American Education* (New Brunswick, N.J.: Rutgers University Press, 1989), 105–122.

15. National Research Council, *Everybody Counts: A Report to the Nation on The Future of Mathematics Education* (Washington, D.C.: National Academy Press, 1989), 2.

Chapter 9. Mathematics of the People, by the People, for the People

1. Jonathan Kozol, *Savage Inequalities: Children in America's Schools* (New York: Crown Publishers, 1991), 127.

2. Paulo Freire, *The Politics of Education* (South Hadley, Mass.: Bergin and Garvey, 1985), 113–116. See also James A. Banks, "The Transformative Approach to Curriculum Reform: Issues and Examples," in *Empowerment through Multicultural Education,* Christine Sleeter, ed. (Albany: The State University of New York Press, 1991), 125–141.

3. Seymour Melman, "Can We Convert from a Military to a Civilian Economy?" *The Churchman's Human Quest,* March, April 1988. Also see Seymour Melman, "The Peace Dividend: What to Do with the Cold War Money," *New York Times,* 17 December 1989, Forum; Leslie H. Gelb, "Foreign Affairs: $1.5 Trillion 'Defense,' " *New York Times,* 17 April 1992, Op-Ed.

4. James S. Jackson, "From Segregation to Diversity: Black Perceptions of Racial Progress," *Rackham Reports (1987–1988)*, Horace H. Rackham School of Graduate Studies, University of Michigan, 55.

5. Quoted in Joanne Rossi Becker and Mary Barnes, "Report on Women and Mathematics: Topic Area at Fifth International Congress on Mathematical Education," *Newsletter* (International Organization of Women in Mathematics Education) 1 (April 1985): 13.

6. Nancy Shelley, "Mathematics: Beyond Good and Evil?" presented at Seventh International Congress on Mathematical Education, 16–23 August 1992.

7. Shelley, "Mathematics."

8. "Briefing: Mathematicians Oppose 'Star Wars.' " *New York Times,* 31 May 1988, B6.

Index

About the Author

Claudia Zaslavsky, a mathematics teacher and consultant, has spent much of her life helping people of all ages and backgrounds get over their fear and enjoy their newfound ability at math. She is the author of *Multicultural Mathematics : Interdisciplinary Cooperative-Learning Activities, Africa Counts: Number and Pattern in African Culture,* and other mathematics-oriented books for children and adults.